D0713057

DATE DUE

AUG 9 1982		
OCT 1 1982		
1985		
WITHDRAWN		

GAYLORD   PRINTED IN U.S.A.

how to
know the

seed
plants

The **Pictured Key Nature Series** has been published since 1944 by the Wm. C. Brown Company. The series was initiated in 1937 by the late Dr. H. E. Jaques, Professor Emeritus of Biology at Iowa Wesleyan University. Dr. Jaques' dedication to the interest of nature lovers in every walk of life has resulted in the prominent place this series fills for all who wonder **"How to Know."**

John F. Bamrick and Edward T. Cawley
Consulting Editors

The Pictured Key Nature Series

How to Know the

AQUATIC INSECTS, Lehmkuhl
AQUATIC PLANTS, Prescott
BEETLES, Arnett-Jaques, Second Edition
BUTTERFLIES, Ehrlich
CACTI, Dawson
EASTERN LAND SNAILS, Burch
ECONOMIC PLANTS, Jaques, Second Edition
FALL FLOWERS, Cuthbert
FERNS, Mickel
FRESHWATER ALGAE, Prescott, Third Edition
FRESHWATER FISHES, Eddy-Underhill, Third Edition
GILLED MUSHROOMS, Smith-Smith
GRASSES, Pohl, Third Edition
IMMATURE INSECTS, Chu
INSECTS, Bland-Jaques, Third Edition
LAND BIRDS, Jaques
LICHENS, Hale, Second Edition
LIVING THINGS, Jacques, Second Edition
MAMMALS, Booth, Third Edition
MARINE ISOPOD CRUSTACEANS, Schultz
MITES AND TICKS, McDaniel
MOSSES AND LIVERWORTS, Conard-Redfearn, Third Edition
NON-GILLED FLESHY FUNGI, Smith-Smith
PLANT FAMILIES, Jaques
POLLEN AND SPORES, Kapp
PROTOZOA, Jahn, Bovee, Jahn, Third Edition

ROCKS AND MINERALS, Helfer
SEAWEEDS, Abbott-Dawson, Second Edition
SEED PLANTS, Cronquist
SPIDERS, Kaston, Third Edition
SPRING FLOWERS, Cuthbert, Second Edition
TREMATODES, Schell
TREES, Miller-Jaques, Third Edition
TRUE BUGS, Slater-Baranowski
WATER BIRDS, Jaques-Ollivier
WEEDS, Wilkinson-Jaques, Third Edition
WESTERN TREES, Baerg, Second Edition

how to
know the
seed
plants

Arthur Cronquist

New York Botanical Garden

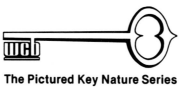

The Pictured Key Nature Series
Wm. C. Brown Company Publishers
Dubuque, Iowa

Contents

Preface

This book is intended to permit the user to identify and learn to recognize the large majority of families of seed plants. Emphasis is placed on plants that grow wild in the United States, or that are frequently cultivated there. The larger or economically important or botanically especially interesting families from outside the United States are also included to round out the picture.

The book is the successor to the part on seed plants in a book by H. E. Jaques, *Plant Families: How To Know Them,* second edition by Wm. C. Brown Company, Publishers, Dubuque, Iowa, 1949. The text has been completely rewritten, but about half of the illustrations are taken from Jaques. Drawings of the following species are taken from *Manual of the Vascular Flora of the Carolinas,* by Albert E. Radford, Harry E. Ahles, and C. Ritchie Bell, and used with the kind permission of the University of North Carolina Press: *Actaea alba, Albizia julibrissin, Bumelia lycioides, Burmannia biflora, Claytonia virginica, Cle-* *thra alnifolia, Cyperus odoratus, Didiplis diandra, Eichhornia crassipes, Eriogonum tomentosum, Euphorbia corollata, Euphorbia maculata, Fraxinus americana, Heuchera americana, Hexastylis virginica, Ilex opaca, Maclura pomifera, Mayaca aubletii, Mirabilis jalapa, Morus rubra, Nymphoides aquaticum, Parnassia asarifolia, Phragmites australis, Piriquetia caroliniana, Platanus occidentalis, Polygonum persicaria, Potamogeton pulcher, Rosa carolina, Ruellia caroliniensis, Ruppia maritima, Saxifraga virginiensis, Schisandra glabra, Scirpus acutus, Smilax rotundifolia, Spergularia marina, Symplocos tinctoria, Tradescantia ohiensis, Urtica dioica, Zannichellia palustris,* and *Zostera marina.* The remainder of the drawings have been prepared especially for this book by Robin A. Jess.

Mabel A. Cronquist, my wife, has been my constant assistant, editorial consultant, and typist. Her help has been essential to the preparation and timely completion of the manuscript.

Introduction

PRINCIPLES OF CLASSIFICATION AND NOMENCLATURE

No two plants are quite alike, any more than two animals are quite alike. Yet the diversity among individuals is not randomly scattered. Certain combinations of characters occur over and over again, with only minor differences among individuals, whereas many other theoretically possible combinations of characters do not occur at all. There are cluster-patterns of all sizes and density in the distribution of diversity, separated by empty or nearly empty spaces of all sizes. In classifying organisms we try to draw lines between groups through the empty spaces rather than through the clusters.

Individuals that are nearly enough alike can usually breed together (if they are not both of the same single sex) and produce fertile offspring. Individuals that are a little more different may be able to breed together, but their offspring are sterile. A mule, for example, is a sterile hybrid between a jackass and a mare. Individuals that are still more different cannot breed together at all. The tale of the man who owned a hybrid between a parrot and a cobra ("He doesn't know what sort of animal it is, but when it talks, he listens!") is a good story but a biological impossibility.

The pattern of breeding in nature tends to correlate with groups that can be recog-

nized by an observant person, and many such groups have been recognized and named in all societies. Variation within such an interbreeding group appears to be continuous, in contrast to the discontinuous variation between groups that do not ordinarily breed together. These recognizable groups of individuals are called species, from the Latin word for a particular ("special") kind.

Sometimes the situation is more complex, and there are breeding barriers between individuals that are otherwise very much alike. Taxonomists differ as to how such groups should be classified. The current view among most animal taxonomists is that intersterile groups should be treated as different species, regardless of how insignificant the other differences may be. Because of a series of features that affect reproduction in plants much more than in animals, this approach is much less satisfactory in plants. Most plant taxonomists consider that there must be a recognizable difference in structure between groups that are classified as different species, even if this means that a species may consist of several reproductively isolated groups.

A working definition of species in plants is that they are the smallest groups that are consistently and persistently distinct, and distinguishable by ordinary means. With a bit of skill and practice, species of plants of ordinary size can usually be recognized in life with the naked eye, or with the aid of a hand lens as one gets a little older and less sharp-eyed. A low-power dissecting microscope may be useful or necessary if one must start from the beginning and examine a flower in detail to make an identification, but the things one is already familiar with can be more readily recognized. Similarly, a detective can recognize his aunt Mary without studying her fingerprints.

The number of species of plants is not accurately known. In temperate regions, where most of the botanists live, we can now make a reasonable approximation of the number of species of vascular plants (seed plants, ferns, and fern-like plants), but our knowledge of many of the fungi and smaller algae is much more primitive. In tropical regions "new" species even of vascular plants continue to be found in considerable numbers, and knowledge of many "known" species is so scanty as to leave much room for doubt. There may be about 350,000 species of plants, of which about 220,000 are angiosperms, or flowering plants, and scarcely more than 500 are gymnosperms. The rest belong to other groups.

The number of species of both plants and animals is so large that a formal scheme of classification is necessary for us to understand and keep track of what we know or believe, and to communicate our thoughts to others. A group of similar species constitutes a genus (plural: genera); a group of similar genera constitutes a family; a group of similar families constitutes an order; a group of similar orders constitutes a class; and a group of similar classes constitutes a division. If more categories are needed, additional ones can be established. Thus a subdivision stands between division and class, a subclass stands between class and order, etc. The division, equivalent to the zoological phylum, is the highest (most inclusive) category that is regularly used in plant classification. With this single exception, the essential categories used in the classification of plants and animals are the same.

The taxonomic system of classification, as here explained, is a hierarchical system. Units of a given rank are assembled into a series of progressively larger groups. A military organization has a similar hierarchy. A number of privates report to a corporal; a number of corporals report to a sergeant; a number of sergeants report to a second lieutenant, and so on up the hierarchy to a general officer. The taxonomic system differs from a military one in that the number of units in each group is more variable. Some genera, such as *Cannabis* and *Ginkgo*, have only a single species each; others have several or even very many species, up to more than a thousand in such genera as *Astragalus, Euphorbia, Senecio,* and *Solanum.*

Species are considered to be the fundamental units of biological classification. All the higher units are groups of species, or groups of such groups, etc. Although these higher units are necessary as aids to understanding and memory, there is nothing inherent about their rank in the hierarchy. A genus is nothing more than a group of species similar enough so that it is useful to think of them collectively; likewise a family is a group of genera that can usefully be considered collectively, and so on.

There is room for a certain amount of unresolvable difference of opinion about the rank of particular groups, especially groups above the rank of species. Thus virtually all botanists are agreed that the plants we call legumes constitute a proper taxonomic group, and that this group consists of three subordinate groups. Some botanists consider that the legumes constitute an order or suborder with three families, the Mimosaceae, Caesalpiniaceae, and Fabaceae. Others recognize a single family Fabaceae (or Leguminosae), with three subfamilies, Mimosoideae, Caesalpinioideae, and Faboideae. Obviously in a formal scheme of classification some choice must be made between such alternatives, but most taxonomists now consider that differences of opinion about the *rank* at which a supraspecific group should be recognized are of relatively small consequence.

The nearest thing to an inherent criterion for the rank of genus is that the species belonging to it should be visually recognizable as a group. Even this criterion can often be met in more than one way, or is subject to differences in interpretation. Still, it is a useful guideline. All the Maples, for example, belong to the genus *Acer,* and all the Oaks to the genus *Quercus.* Our colonial forebears, familiar with Maples and Oaks in Europe, had no difficulty in recognizing other species of these genera in America. They also quickly came to recognize another group of trees, unknown in Europe, for which they adopted the Indian name Hickory. The several species of Hickory are now classified as forming a genus *Carya,* a name slightly Latinized from the Indian. Not all genera are so readily recognizable as these three, but the concept of visually recognizable genera is still useful.

Scientific nomenclature of plants and animals is in Latin, or Latin form. Originally this was because Latin was the language of learning. People who wrote scholarly works customarily wrote in Latin, so naturally the names were in Latin. Latin has been retained for scientific names because it has certain advantages.

First, Latin is an international language, inherited from the past, which is available for use by all without injury to national pride. Thus the same kind of plant or animal has the same correct scientific name in all countries, regardless of language differences. The generic name *Quercus* is the same throughout the world, although the vernacular name for it in English is Oak, in French Chêne, in German Eiche, and in Russian дуб. *Quercus*

robur, which we know as English Oak, grows throughout much of Europe, with different common names in different countries, but scientifically it is still *Quercus robur* wherever it grows.

Second, the use of Latin gives us a chance to impose some order on scientific nomenclature. The name Buttercup may mean very different things to people in Pennsylvania, Texas, and Idaho. The same garden flower may be known to one person as Four-O'Clock, and to another as Marvel of Peru. An Eastern American tree known to many people as Tulip Poplar is not related to either Tulips or Poplars. Since it is usage that makes correctness in language, these names as applied are not "wrong," merely confusing. For scientific names, on the other hand, we can and do have an agreed set of rules for choosing the correct name, once the taxonomic position and status of a plant or animal are determined. These rules do not work perfectly, any more than the laws of a country work perfectly, but they provide the basis for order and ultimate uniformity.

Scientific nomenclature follows a binomial system; that is, the name of a particular species of plant or animal consists of two words. The first word is the name of the genus, and the second word indicates the particular species within the genus. For example, all maples belong to the genus *Acer.* The generic name *Acer* can stand alone to indicate any or all species of Maple. The Red Maple is *Acer rubrum,* the Black Maple is *Acer nigrum,* and the Striped Maple is *Acer pensylvanicum.* In association with the name *Acer,* the words *rubrum, nigrum,* and *pensylvanicum* are specific epithets that indicate particular kinds of Maple. By themselves the words *rubrum, nigrum,* and *pensylvanicum* are merely Latin adjectives meaning red, black, and of Pennsylvania, respectively. The same adjectives can be used again as specific epithets in other genera. There is an *Allium ru-*

brum, for a species of Onion, a *Solanum nigrum,* for Black Nightshade, and a *Saxifraga pensylvanica,* for a species of Saxifrage.

The binomial system of scientific nomenclature is a formalized adaptation of a widespread but more casual binomial system of folk nomenclature. The names Red Maple, Black Maple, and Striped Maple, and Scarlet Oak, Pin Oak, and Post Oak are just as truly binomials as the corresponding scientific names, *Acer rubrum, Acer nigrum, Acer pensylvanicum, Quercus coccinea, Quercus palustris,* and *Quercus stellata.* In English the generic name is the second of the two words, rather than the first, but in some other languages it is the other way around as it is in scientific nomenclature.

The binomial system of nomenclature was established by the Swedish naturalist Carl Linnaeus. Prior to Linnaeus' time most scholarly names of species consisted of several words, essentially a short description. The idea of a binomial system did not originate with Linnaeus, but he applied it uniformly in a way that led to its rapid acceptance. By agreement among botanists, scientific botanical nomenclature begins in 1753, with the publication of Linnaeus' massive work, Species Plantarum, in which he tried to list and classify all the plants in the world.

In formal scientific nomenclature it is customary to add the name of the author (or an abbreviation of it) to the Latin name. This practice is partly a bibliographic device, to help locate the source of the name, but mainly it is a means of reducing the possibility of confusion. It sometimes happens that the same name is proposed independently by two authors for two different things, the second author being unaware of the earlier name. When such an instance comes to light, the later name must ordinarily be dropped, even though the earlier name may have been presented in an obscure publication. The routine citation of the name of the author of a

species helps to keep track of things. Thus, although the name *Acer rubrum* is technically correct and sufficient for the Red Maple, and *Acer nigrum* for the Black Maple, it is customary to write *Acer rubrum* L. and *Acer nigrum* Michx.f. The L. stands for Linnaeus, and the Michx.f. for Francois André Michaux, son (f.) of another well known botanist, André Michaux.

In botanical writing it is customary to cite both the original author and the correcting author, when a name has been modified since its original publication. Thus the botanical name of the Shepherd's Purse is commonly written *Capsella bursa-pastoris* (L.) Medic. Linnaeus originally published the name as *Thlaspi bursa-pastoris,* but Medicus established the new genus *Capsella* for this species, and called it *Capsella bursa-pastoris.*

Every species of plant or animal has its own ecological niche, or way of life to which it must be morphologically and physiologically adapted. Individuals compete with others of their species, and with other organisms in general, to survive and reproduce. In a more abstract sense, different species compete for the available ecological niches. Among the higher animals, in particular, this interspecific competition commonly leads fairly evidently to competitive exclusion: Only one species occupies a particular ecological niche at a given time and place. Although plant species certainly compete, the competition among them often does not lead to such an obvious degree of competitive exclusion. Frequently several different species appear to be exploiting virtually the same ecological opportunity.

Most kinds of animals must move about to get their food (and often also to keep from being used as food), and most of them have a fairly limited range of acceptable or useful foods. Even the so-called omnivores do not in fact eat everything. The stems and leaves of grasses, which provide the mainstay for many grazing animals, are virtually useless to most other kinds of vertebrates. The whole structure and physiology of an animal must be well adapted to getting more than a fair share of a limited supply of food. Everything must mesh with everything else, and there is little room for casual variation. Once an evolutionary line adopts some general way of making a living, the pressure is toward ever closer adaptation to that way, and toward partitioning the general way of life into a set of more refined adaptive niches. Evolutionary opportunities to move from one general way of life to another are rare. We cannot expect that a group of grazing animals will ever evolve from members of the cat family.

The evolutionary processes among animals tend to produce a taxonomic pattern of well defined groups that are strongly correlated with way of life. Such groups are often evident even to people without formal training in zoology, as is shown by the well established common names. The vertebrates constitute a major subphylum of the phylum Chordata. The five major taxonomic groups of vertebrates are all well known to the general public under the names fish, amphibians, reptiles, birds and mammals. Many of the orders and families of mammals have well established common names. Cats and dogs, for example, belong to different families of the order Carnivora (carnivores).

The evolutionary patterns among higher plants are significantly different from those among higher animals. Plants mostly remain rooted to a particular spot. They mostly make their own food, using essentially the same raw materials and the same source of energy (sunlight). They do not have and do not need the complex sets of adaptations that permit animals to move about and use different kinds of food. Their less closely structured growth system, coupled with a fixed physical position, permits more casual variation in structure without significant harm. Adaptations to particular habitats have a simpler basis, so that

evolutionary change from one habitat to another is not so difficult. Competitive exclusion does not work nearly so efficiently, and plants of rather different appearance can play similar ecological roles.

In conformity to the difference in evolutionary pattern, the taxonomic pattern among higher plants is significantly different from that among higher animals. Every species must of course be suited to its habitat, and there are still concentrations and empty spaces in the distribution of diversity, but the ecological significance of many of the higher taxonomic groups is minimal. Many individual families, such as the Asteraceae, Euphorbiaceae, and Fabaceae include plants of very diverse appearance and occur in a wide range of habitats.

The number of families and orders that are readily recognizable to all as part of our common social heritage is very limited. The most immediately notable such groups are the Palm family (Arecaceae) and the Cactus family (Cactaceae), although many African species of Euphorbiaceae would be taken for cacti by the uninitiated. The Grass family (Poaceae), which might be thought to provide another example, would actually be confused with such families as the Cyperaceae, Juncaceae, and Restionaceae by people with no special training in botany. Extant English names for other families, such as the Orchid family, Heath family, Lily family, and Mustard family, are fostered more by scholarly effort than by genuine public recognition.

In contrast to the families and orders, the number of genera of plants that have genuine common names is fairly large. Many people can recognize an *Iris* or a *Gladiolus,* but few among the lay public know that both genera belong to the Iridaceae. The genera *Quercus* and *Fagus* are well known under the names Oak and Beech, but their membership in the family Fagaceae is known only to the more learned. Although the adaptive significance of the features that characterize a genus of plants is often obscure, the morphological pattern is generally sufficiently constant to permit recognition of the group. There is a degree of circularity here, because taxonomists are reluctant to put things into the same genus if they look very different.

With or without an emphasis on adaptive significance, taxonomy proceeds in any case by recognition of multiple correlations. An effort is made to group together the things that are most alike in all respects, and to separate these groups progressively in the taxonomic system from things they are progressively less like. When the groups have been recognized and delimited on the basis of all the available information, the most constant (or most nearly constant) differences among them are noted and become the important characters. The taxonomic value of individual characters is determined *a posteriori,* after the groups have been recognized, rather than being used beforehand, *a priori,* to establish the groups.

One of the important principles that has been discovered by the foregoing method is that value of a particular character varies from group to group. Nearly all members of the order Gentianales have opposite leaves. Within the order Lamiales, the family Boraginaceae nearly always has alternate leaves, whereas the families Lamiaceae and Verbenaceae nearly always have opposite or whorled leaves. Within the family Scrophulariaceae, many of the genera have opposite leaves, and others have alternate leaves. Within the genus *Cornus,* in the family Cornaceae, the single species *Cornus alternifolia* has alternate leaves, in contrast to the opposite leaves of other species. The species *Helianthus annuus* L. (Asteraceae) and *Cannabis sativa* L. (Cannabaceae) commonly have the lower leaves opposite and the upper leaves alternate.

Yet the taxonomic value of characters is not completely helter-skelter in different groups. On the basis of the experience of generations of taxonomists over the past 200 and more years, it has turned out that characters of the structure of the flower are more likely to be useful in distinguishing families and orders than are most vegetative characters. Among floral characters, the number and arrangement of parts are more likely to be important than their size. Color of the flowers is seldom useful at a higher level than the genus, and the difference between yellow and the anthocyanin colors is more likely to be important than the differences in shades of anthocyanin (blue to purple and crimson). We can be reasonably confident in any particular case that the difference between an inferior and a superior ovary will turn out to be more important than a difference in the shape of the leaves, even though the single genus *Saxifraga* shows a complete range from species with a superior ovary to species with an inferior ovary.

Past experience provides some basis for evaluating the prospective significance of a particular character in the next instance, but the proof of the pudding is always in the eating. Taxonomic groups are still properly recognized on the basis of all the available information, rather than being arbitrarily organized on preconceived characters. There is just enough consistency in the value of characters from group to group to mislead successive generations of taxonomic tyros to suppose that the characters have an inherent, fixed importance, and just enough inconsistency to lead to absurdity in particular cases if such constancy is assumed.

THE CLASSIFICATION OF SEED PLANTS

This book treats the plants that produce seeds as part of their normal reproductive cycle. Botanists agree that except for some ancient fossils the seed plants form two major groups. The English names for these groups are gymnosperms and angiosperms; the angiosperms are also called flowering plants. In the past it was customary to treat all the seed plants as a single division, Spermatophyta, and to recognize the gymnosperms and angiosperms (under appropriate technical names) as subdivisions or classes. Now it is becoming more customary to treat the gymnosperms and angiosperms as separate divisions. The formal nomenclature is still in a state of flux, but there is some body of precedent for the technical names here used, Pinophyta for the gymnosperms, and Magnoliophyta for the angiosperms. It cannot truly be said that the present scheme of two divisions for seed plants is "right" and the earlier scheme of a single division is "wrong." This is a matter on which taxonomists may legitimately differ.

Although the gymnosperms dominated the land vegetation of the earth in the distant past, notably in the first two-thirds of the Mesozoic era, they are now a relic group. Only a few hundred species remain. There has been so much extinction that the surviving groups are easily sorted into a number of well marked families, orders, classes and subdivisions. Although the species are relatively few in number, some of them are abundant and give character to the landscape especially in cool-temperate regions.

The angiosperms are taxonomically much more complex than the gymnosperms. New species and genera are still actively evolving in many of the families, and the limits even between the higher taxa are often difficult to discern. All of the characters used to mark major groups are subject to exception. A genus may obviously belong with a certain family on the totality of its features, and yet differ in some one feature that otherwise helps to distinguish the family from others. Individuals of a species, or different parts on the same individual, may also vary for no obvious rea-son. Students soon learn that it is wise to look at more than one flower before proceeding with identification.

Botanists have agreed for many years that the angiosperms fall naturally into two major groups. These groups are usually called dicotyledons and monocotyledons (or dicots and monocots), from the most nearly constant difference between them. In this book they are treated as classes, under the technical names Magnoliopsida (dicotyledons) and Liliopsida (monocotyledons). There are exceptions to all the differences between dicots and monocots, but the whole ensemble of features seldom leaves any doubt as to the position of a particular species in one group or the other. With a little experience, one can learn to recognize members of these two groups fairly easily. The evolutionary divergence of monocots from dicots evidently took place in the Lower Cretaceous period, not long after the origin of the angiosperms from some group of gymnosperms.

The orders and subclasses of angiosperms are much less clearly marked than the classes or the families. At the present state of knowledge, the orders and subclasses are more useful to the professional taxonomist than to the person starting out to learn the flora of a region. No effort is made in this book to characterize the orders and subclasses of angiosperms, although they are included in the outline of classification at the back.

Most plant taxonomists now agree that the thousands of genera of angiosperms can usefully be organized into some 350 to 400 families. The exact number depends mainly on the degree of splitting or lumping, such as whether one recognizes one family of legumes or three. Aside from the problems of splitting or lumping, only a relatively few genera are booted about from one family to another by different taxonomists. The vast majority of genera have found family homes from which no one wants to uproot them. At the same time, one must recognize that most of the larger families include one or more genera that are aberrant and do not have all the usual features of the group. It is impossible to make a key that will provide for all members of all families, and also bring all members of each family out in only one place.

Although some 375 families of angiosperms are recognized in this book, about half of them are relatively small, with not more than a hundred species each. Furthermore, many of the families do not occur in the United States. In the keys that follow, an effort is made to include all families that are represented in the wild in the continental United States or are frequently cultivated there, and also all families that are economically important or are of especial botanical interest. In effect this means that nearly all the families with as many as a hundred species are included in the keys, as well as some of the smaller families.

Some compromise has been necessary between the effort to provide for all members of a family and the effort to bring out all members of a family at the same place in the key. Inconspicuous exceptions to the standard family characters, or exceptions that do not involve plants of the United States, are mostly ignored. Thus no effort is made to provide for the genus *Subularia*, a submersed aquatic member of the Brassicaceae that is rarely noted except by determined botanists. Likewise the small genus *Besseya*, in the Scrophulariaceae, which lacks a corolla, is ignored. On the other hand, a special entry is made for *Prunus*, a familiar genus of the Rosaceae that differs from most members of its family in having flowers with a single carpel.

Although the total number of families of angiosperms may seem formidable, the number of really large families is relatively limited. There are only some 33 families with 2000 or more species, and several of these are mainly tropical. In the following list of the largest families, the approximate number of species is given in parentheses after each one, and the families that are insignificant in the continental United States are marked with an asterisk:

> Orchidaceae (20,000), Asteraceae (20,000), Fabaceae (10,000), Poaceae (8000), Euphorbiaceae (7,500), Rubiaceae (6500), Liliaceae (4000), Cyperaceae (4000), *Melastomataceae (4000), Scrophulariaceae (4000), Lamiaceae (3200), *Arecaceae (3000), Apiaceae (3000), Rosaceae (3000), *Myrtaceae (3000), Brassicaceae (3000), Verbenaceae (2600), *Acanthaceae (2500), Ericaceae (2500), *Aizoaceae (2500), Solanaceae (2300), *Annonaceae (2300), Caesalpiniaceae (2200), Mimosaceae (2000), Apocynaceae (2000), Asclepiadaceae (2000), Boraginaceae (2000), Ranunculaceae (2000), Campanulaceae (2000), Cactaceae (2000), Caryophyllaceae (2000), *Gesneriaceae (2000), and *Lauraceae (2000).

Another 10 families, although not so large as those listed above, are very common in the United States, either in terms of number of species or in terms of abundance of individuals that will attract attention. These are:

> Aceraceae, Chenopodiaceae, Fagaceae, Hydrophyllaceae, Juglandaceae, Onagraceae, Polemoniaceae, Polygonaceae, Salicaceae, and Saxifragaceae.

Thus the list of the most abundant and conspicuous families in the continental United States can still be kept to about thirty-five. Learn to recognize these without using the keys, and you will know the family of the vast majority of the flowering plants you see growing wild in the United States.

How to Use the Keys

A botanical key is an arrangement of information that is intended to permit easy identification without requiring the user to read all about every group. Most keys are dichotomous, that is they always provide just two choices. They work something like the old parlor-game of twenty questions. Each choice between two alternatives in the key eliminates all the things under the alternative not chosen. Each selection of one alternative leads to another choice between two alternatives, until eventually the answer is found, the plant identified.

There are several ways of organizing the choices in a key on the printed page. In this book, each opportunity for choice is given a number, and each of the two choices is distinguished by a letter (either a or b). The choice, once made, leads to another number at the right-hand margin of the page. Go to that number, and you will again find two choices, a and b, and so on until the identification is made.

Key to the Families of Seed Plants

1a Ovules and seeds borne on the surface of bracts or scales, these often arranged in cones, or the ovules and seeds borne at the end of a short or long stalk, in any case the ovule exposed to the air at the time of pollination, the pollen landing at the micropyle; male (pollen-bearing) cones always separate from the female, often soft and short-lived; woody plants (seldom with the stem so short that the leaves appear to arise directly from the ground). (Division I, PINOPHYTA, the Gymnosperms) 2

1b Ovules and seeds enclosed within an ovary, the seeds released (if at all) only at maturity; pollen landing on a specialized structure, the stigma, which is typically elevated above the ovary on a slender style; ovary (containing the ovules) and stamens (producing the pollen) borne in the same flower or in different flowers; plants woody or herbaceous. (Division II, MAGNOLIOPHYTA, the Angiosperms) 11

DIVISION PINOPHYTA
The Gymnosperms

2a (from 1a) Leaves pinnately compound, large and fern-like, evergreen, clustered atop a simple, unbranched trunk, or arising from the surface of the ground. (Subdivision Cycadicae) ....
.... Cycad Family, CYCADACEAE

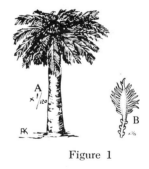

Figure 1

Figure 1 *Cycas revoluta* Thunb., Sago Palm. A, mature plant; B, megasporophyll (fertile leaf) with ovules.

The Sago Palm is not a true palm, but a cycad. Cycads often resemble palms (see fig. 260) in general appearance, but they are very different in anatomical and reproductive features. In *Cycas* the megasporophylls stand out separately at the top of the stem. In other cycads, such as *Zamia floridana* DC., native to Florida, the megasporophylls are smaller and arranged into cones that look something like large pine cones. The microsporophylls (pollen-bearing leaves) of all cycads are borne in cones.

There are fewer than a hundred living species of cycads, all essentially tropical. They are the dwindling remnant of a group of plants that were abundant and diversified in the Mesozoic Era.

2b   Leaves simple; branching trees or shrubs, except Welwitschiaceae **3**

3a   (from 2b) Leaves opposite, or in whorls of three, never so closely crowded as to cover the surface of the stem; wood with vessels but without resin canals; male cones compound; embryo of the seed with two cotyledons. (Subdivision Gneticae) **4**

3b   Leaves borne singly (more or less spirally), or in clusters on very short spur-branches, or sometimes in opposite pairs or whorls of three, in the latter case small, scale-like, and closely crowded, largely covering the surface of the twigs; wood without vessels but usually with resin canals; male cones simple; embryo of the seed with several cotyledons (but these sometimes scarcely developed when the seed first appears to be ripe). (Subdivision Pinicae) **6**

4a   (from 3a) Leaves tiny, scale-like, in widely separated, opposite pairs or whorls of three; much-branched shrubs with jointed, photosynthetic twigs .... **...... Ephedra Family, EPHEDRACEAE**

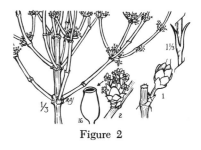

Figure 2

Figure 2 *Ephedra nevadensis* S. Wats., Mormon Tea, showing branching stem with tiny leaves; male cone, with an enlarged microsporangium drawn separately; female cone, with two ovules protruding; portion of young twig, with tiny, opposite leaves.

The family Ephedraceae has only the genus *Ephedra,* with about 40 species, native to deserts in both the Old and the New World. All species produce the well-known drug ephedrine. The stems and cones have been used in pioneer communities to brew a tonic beverage. Male and female cones of *Ephedra* are borne on separate plants. Both kinds of cones are small, seldom more than about one cm long. The female cones have one to three seeds that protrude beyond the small cone-scales.

4b   Leaves well developed, of ordinary or even large size **5**

5a   (from 4b) Leaves numerous, net-veined, resembling the leaves of dicotyledons; trees, shrubs or vines, evidently branched ...... **Gnetum Family, GNETACEAE**

Figure 3

Figure 3 *Gnetum indicum* (Lour.) Merrill, a species of southern Asia and the East Indies; habit of male plant, with an enlarged male cone at the lower right, and young female cone at upper right.

The 30 or so species of *Gnetum* are all tropical. *Gnetum, Ephedra,* and *Welwitschia,* each the only genus of its family and order, resemble angiosperms in having vessels (specialized water-conducting tubes) in the wood, but all three genera have gymnospermous reproductive structures. The leaves of *Gnetum* look very much like angiosperm leaves, and *Gnetum gnemon* L., in particular, is remarkably similar in aspect to *Coffaea arabica* L., the coffee plant, a member of the family Rubiaceae.

5b Leaves only two, large (sometimes more than two m long), strap-shaped, spreading out on the ground away from the short, broad, unbranched stem, persistent and eventually becoming frayed and disheveled .... Welwitschia Family, WELWITSCHIACEAE

Figure 4

Figure 4 *Welwitschia mirabilis* Hook.f.; habit, with progressively enlarged details of male cones. *Welwitschia* is one of the most bizarre plants in the world, found only in deserts in southwestern Africa. The plants have a very long taproot and attain an age of more than 100 years. The saucer-shaped or top-shaped stem is short and broad, sometimes becoming more than one m thick. The two leaves grow indefinitely from the base and die back from the tip. Like *Gnetum* and *Ephedra, Welwitschia* is dioecious; i.e., the pollen and seeds are produced on different plants.

The genus has only a single species.

6a (from 3b) Seeds borne singly or paired on a stalk, fully exposed, not in cones, provided with a fleshy seed-coat or aril 7

6b Seeds borne in cones (these sometimes of ordinary appearance, or sometimes small, pea-sized, and resembling a firm, dry berry, but the cone-scales still discernible); seed-coat firm and dry, without an aril; the families of this group, making up the order Pinales, are collectively called conifers 8

7a (from 6a) Leaves deciduous, fan-shaped, often bilobed, with numerous forking (dichotomous) veins; trees ....
.... Ginkgo Family, GINKGOACEAE

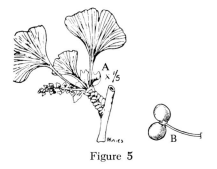

Figure 5

Figure 5 *Ginkgo biloba* L., Maidenhair Tree; A, twig with leaves and male cones on a dwarf branch; B, a pair of ripe seeds on their common stalk.

The Maidenhair Tree is the sole living species of its family, order, and class, the last remnant of an ancient lineage that originated during the Paleozoic Era. Not now known to grow in a native, wild condition, it has long been cultivated in China and is occasionally grown as a street tree in the United States. The flesh of the plum-like seeds of the female tree has a foul odor, but the nut-like central kernel is esteemed as a delicacy in the Orient.

7b   **Leaves evergreen, narrow, linear, with a single midvein; trees and shrubs .... Yew Family, TAXACEAE**

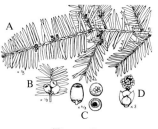

Figure 6

Figure 6 *Taxus baccata* L., English Yew; A, twig with needle-like, evergreen leaves, and male cones; B, portion of twig with berry-like seeds; C, side, bottom and top views of seed; D, male cone.

Yews have glossy, flat needles that spread out in two rows, one row on each side of the twig. The needles are highly poisonous if eaten, the fleshy, bright orange-red seeds somewhat less so. The eastern American Yew (*Taxus canadensis* Marsh.) is a straggling shrub, but the English Yew, often planted for ornament, will grow into a dense, handsome, small tree that can be trimmed into various shapes. Oregon Yew (*Taxus brevifolia* Nutt.) is favored by archers as wood for making bows. There are about 20 species of Taxaceae.

8a   **(from 6b) Each cone-scale bearing a single ovule and thus only one seed; leaves at least three mm wide at the base .... Araucaria Family, ARAUCARIACEAE**

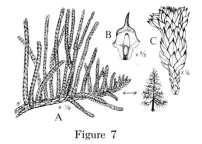

Figure 7

Figure 7 A, *Araucaria excelsa* R. Br., Norfolk-Island Pine, with cone-scale, B, showing the single seed; C, *Araucaria araucarana* (Mol.) C. Koch, Monkey Puzzle.

There are about 40 species of Araucariaceae, native chiefly to the Southern Hemisphere. A few species are planted as ornamental trees in the United States. The Norfolk-Island Pine, not a true pine, becomes a noble tree, as much as 60 m. tall. The seed of *Araucaria* is so closely grown to the cone-scale that it is often difficult to tell where the seed stops and the cone-scale begins.

8b   Each cone-scale bearing two (-several) ovules and thus potentially two (-several) seeds; leaves less than three mm wide at the base **9**

9a   (from 8b) Cone-scales (of the female cones) opposite or in whorls of three; leaves needle-like, or very often scale-like and clothing the twigs; cones of ordinary appearance, or sometimes small and berry-like .... **Cypress Family, CUPRESSACEAE**

Figure 8

Figure 8   A, *Juniperus virginiana* L., Eastern Red Cedar, twigs and berry-like cones; B, *Thuja plicata* Donn, Western Red Cedar, twigs and cones.

There are about 130 species of Cupressaceae. Species of *Cupressus, Chamaecyparis, Juniperus* and *Thuja* are often planted for ornament. *Thuja plicata* has soft, aromatic wood that is resistant to decay and makes excellent shingles. It is an important timber tree in the Pacific Northwest, where in the absence of fire it tends to replace the thick-barked, more fire-resistant Douglas Fir.

9b   Cone-scales (of the female cones) spirally arranged; cones not berry-like **10**

10a   (from 9b) Principal cone-scales each subtended by a bract (but this often hidden by the scales next below in the cone); leaves needle-like .... **Pine Family, PINACEAE**

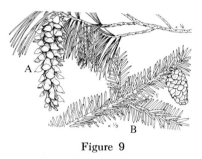

Figure 9

Figure 9   A, *Pinus strobus* L., White Pine, twig with needles and cone; B, *Picea glauca* (Moench) Voss, White Spruce, twig with needles and cone.

There are about 200 species of Pinaceae, most of them in North Temperate regions, many of them important timber or ornamental trees. *Abies* (Fir), *Larix* (Larch, with deciduous leaves), *Picea* (Spruce), *Pinus* (Pine), *Pseudotsuga* (*P. menziesii* is the Douglas Fir), and *Tsuga* (Hemlock) are the most important genera. The boreal forest in Canada, Alaska, and northern Eurasia is dominated by members of the Pinaceae, especially Spruce and Fir. Farther south, *Tsuga canadensis* grows intermingled with deciduous trees (in northeastern U.S.), and several species of pine are known as fire-trees because of their resistance to fire or their ability to regenerate rapidly after fire.

10b   Principal cone-scales not subtended by bracts; leaves needle-like or sometimes scale-like .... **Bald Cypress Family, TAXODIACEAE**

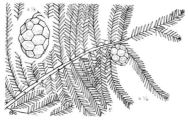

Figure 10

Figure 10 *Taxodium distichum* (L.) Richard, Bald Cypress, twig and cone. Bald Cypress is unusual among the conifers in that the leafy twigs are deciduous in the autumn, so that the tree is bare ("bald") in the winter. The Dawn Redwood, *Metasequoia*, resembles Bald Cypress in this respect, but other members of the family, such as the California Big Tree (*Sequoiadendron*) and the Coast Redwood (*Sequoia*) are evergreen.

There are only about 16 species of Taxodiaceae. The family includes the tallest (*Sequoia sempervirens*) and the most massive (*Sequoiadendron giganteum*) trees in the world, but not the oldest (*Pinus longaeva*, family Pinaceae, in the Intermountain region of Western United States).

DIVISION MAGNOLIOPHYTA
The Angiosperms

11a   (from 1b) Leaves usually net-veined (A); flower parts typically in sets of five or four (B); stems usually either herbaceous and with the vascular bundles in a ring surrounding the pith (C), or woody and increasing in thickness year by year, commonly forming annual growth-rings; embryo of the seed usually with two cotyledons (Class Magnoliopsida, the Dicotyledons) 12

11b   Leaves usually parallel-veined (A); flower parts typically in sets of three (B); stems typically herbaceous and with scattered vascular bundles (C), though sometimes becoming hollow at maturity, or sometimes woody but then not increasing much if at all in thickness over the years, and not forming annual growth-rings; embryo of the seed with only one cotyledon, or the embryo not divided into parts (Class Liliopsida, the Monocotyledons) 250

Class Magnoliopsida, The Dicotyledons

12a   (from 11a) Plants insectivorous 13

12b   Plants not insectivorous 16

13a   (from 12a) Corolla sympetalous, irregular, two-lipped and more or less distinctly five-lobed, the lower lip sometimes with a spur; ovary unilocular, with basal or free-central placentation .... Bladderwort Family, LENTIBULARIACEAE

Figure 11

Figure 11 A, *Utricularia vulgaris* L., Greater Bladderwort; B, *Pinguicula vulgaris* L., Butterwort.

There are about 200 species of Lentibulariaceae, most of them in the genera *Utricularia* and *Pinguicula*. Bladderworts (*Utricularia*) mostly float in shallow water, and have showy yellow flowers that rise a few inches above the surface. The leaves are generally wholly submerged and dissected into thread-like segments, some of which bear small bladders that trap tiny aquatic animals. Butterworts (*Pinguicula*) have purple or yellow flowers with an evident spur. The plants grow on damp ground and have sticky leaves that act like fly-paper, trapping insects that are then at least partly digested. The Lentibulariaceae are related to the large and widespread family Scrophulariaceae.

13b **Corolla of separate petals, or wanting** **14**

14a **(from 13b) Leaves, or some of them, modified to form pitchers; ovary with 3-5 locules and axile placentation 15**

14b **Leaves not forming pitchers; ovary with a single locule and parietal or basal placentation; plants of bogs and similar wet, low places .... Sundew Family, DROSERACEAE**

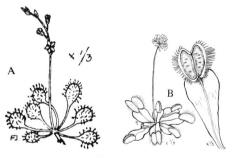

Figure 12

Figure 12 A, *Drosera rotundifolia* L., Round-leaved Sundew; B, *Dionaea muscipula* Ellis, Venus' Flytrap.

The hundred species of Droseraceae are mostly small plants with a rosette of leaves at the ground-level, and with a short, erect flowering stalk. The leaves of Sundew (*Drosera*, by far the largest genus) are covered with irritable, mucilage-tipped tentacle-hairs, which bend and hold the captured insect. The leaves of Venus' Flytrap (one local species of southeastern U.S.) are hinged along the midrib; the leaf snaps shut when two or more of its several trigger-hairs are touched. Some botanists consider that the Droseraceae, Sarraceniaceae, and Nepenthaceae are fairly closely related and should collectively be treated as an order Nepenthales. Others would dissociate one or another of the three families from the other two, or even distribute them among three different orders.

15a **(from 14a) Flowers perfect; filaments distinct; bog plants, not climbing; New World plants .... Pitcher Plant Family, SARRACENIACEAE**

Figure 13

Figure 13 A, *Sarracenia purpurea* L., Northern Pitcher Plant; B, *Darlingtonia californica* Torr., Cobra Plant or California Pitcher Plant, with a portion of a flower stalk shown at somewhat larger scale than the leaves.

There are only three genera of Sarraceniaceae, two in the United States, the third (*Heliamphora*) in northern South America. Leaves of pitcher plants have stiff, smooth, downward-pointing hairs on the inner side near the top. Further down they are smooth and partly filled with digestive liquid. The several species of *Sarracenia* are commonest on the coastal plains of southeastern United States; they have different odors and attract differents kinds of insects. The Cobra Plant grows near the coast in southern Oregon and northern California.

15b Flowers unisexual; filaments united into a column; tropical forest herbs or more often shrubs or half-shrubs, often climbing or epiphytic; Old World tropical plants .... Old World Pitcher Plant Family, NEPENTHACEAE

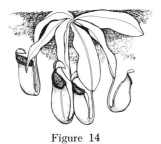

Figure 14

Figure 14 *Nepenthes sanguinea* Lindl., Malayan Pitcher Plant.

Nepenthes, with about 75 species, is the only genus of Nepenthaceae. It has complex leaves, with a short petiole and a flattened blade. The blade narrows at the tip into a stout tendril that connects to a large, terminal pitcher. Some species are grown as curiosities in greenhouses.

16a (from 12b) Plants either without chlorophyll (and then with tiny or no leaves), or obviously parasitic and attached to the branches of trees 17

16b Plants with chlorophyll and usually (not always) with well developed leaves, not parasitic, or not obviously so (this group

includes some green-leaved root-parasites, as well as many ordinary autotrophs) 23

**17a (from 16a) Plants slender, twining, non-green, with threadlike or finely cordlike stems; flowers small, perfect, regular, sympetalous; stamens as many as the corolla-lobes and alternate with them; corolla-tube with a set of variously fringed or cleft scales directly beneath the stamens ....
.... Dodder Family, CUSCUTACEAE**

Figure 15

Figure 15 *Cuscuta gronovii* Willd., Dodder. There are about 150 species of Dodder, found throughout most of the world. Dodders are closely related to the Morning Glories (Convolvulaceae), which often have slender, twining stems, but which have green leaves and are not parasitic. Dodder seeds germinate on the ground and develop a small root-system, but the basal parts wither away after the stem has made contact with the host plant and developed haustoria.

The tropical genus *Cassytha* (Love-vine) is an unusual member of the family Lauraceae that resembles *Cuscuta* in appearance and way of life. The chiefly West Indian species *Cassytha filiformis* L. also occurs in southern Florida. *Cassytha* is easily distinguished from *Cuscuta* by its perianth of six separate sepals, and by having nine stamens and three staminodes.

17b Plants otherwise, not twining, and not with notably slender stems **18**

18a (from 17b) Plants more or less green and chlorophyllous, seated on the branches of the host tree; leaves opposite, sometimes much reduced; ovary inferior **19**

18b Plants not green, lacking chlorophyll, seated in the ground **20**

19a (from 18a) Flowers perfect, and generally with a more or less well developed calyx in addition to the showy corolla .... **Flowering Mistletoe Family, LORANTHACEAE**

Figure 16

Figure 16 *Psittacanthus calyculatus* (DC.) G. Don, Flowering Mistletoe, a native of Mexico. A, habit of flowering plant; B, fruits. There are about 900 species of Loranthaceae, called Flowering Mistletoe because their flowers are so much larger and more showy than those of the more familiar Mistletoes of temperate regions. Visitors to tropical parts of the New World may expect to see *Psittacanthus* growing from the limbs of trees.

19b Flowers unisexual and with a single set of tepals, generally small and inconspicuous .... **Mistletoe Family, VISCACEAE**

Figure 17

Figure 17 A, *Phoradendron serotinum* (Raf.) M. C. Johnst., American Mistletoe; B, *Arceuthobium americanum* Nutt., Dwarf Mistletoe.

American Mistletoe grows on many kinds of trees throughout the United States, and is the common Christmas Mistletoe in this country. Dwarf Mistletoe has tiny, scale-like leaves; it grows on several kinds of Pine in western United States.

There are about 300 species of Viscaceae, found throughout most of the world. The Viscaceae are evidently related to the Loranthaceae, and many botanists especially in the past have included both groups in a more broadly defined family Loranthaceae.

20a (from 18b) Ovary inferior or half-inferior; filaments united into a tube surrounding the style, or more often adnate to the style; flowers with a single set of tepals, often showy and large or very large, borne singly at the ground level, or several on a short, fleshy spike; plants parasitic .... **Rafflesia Family, RAFFLESIACEAE**

Figure 18

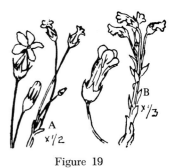

Figure 19

Figure 18 *Rafflesia arnoldii* R. Br., Giant Rafflesia. This species, native to Sumatra, is famous for having the largest flower in the world, sometimes fully a meter across.

Members of the Rafflesiaceae are mainly tropical, and none of the 50 species is native to the United States. The vegetative body of members of the Rafflesiaceae is filamentous and fungus-like, permeating the tissues of the host. The affinities of the Rafflesiaceae are in dispute.

20b **Ovary superior; filaments distinct, free from the style; leaves alternate, very small; flowers with both sepals and petals** **21**

21a **(from 20b) Flowers irregular, the petals united to form a two-lipped, five-lobed corolla; functional stamens four (an additional staminode sometimes present); ovary unilocular, with (2) 4 (6) intruded parietal placentas; ovules numerous; root-parasites (a few parasitic Scrophulariaceae might be sought here, except that the ovary is bilocular, with axile placentation)** .... **Broom-Rape Family, OROBANCHACEAE**

Figure 19 A, *Orobanche uniflora* L., Cancer-Root; B, *Orobanche fasciculata* Nutt., Yellow Cancer-Root.

There are about 150 species of Orobanchaceae, 100 of them in the genus *Orobanche*. The Orobanchaceae are closely related to the large family Scrophulariaceae. Many Scrophulariaceae are partly parasitic on the roots of other plants, but still have green leaves and make much of their own food. Only a few Scrophulariaceae, such as *Striga*, have reached the same level of parasitism as the Orobanchaceae and become completely dependent on their hosts, without photosynthetic pigments.

21b **Flowers regular or nearly so; petals separate or united to form a sympetalous corolla; stamens as many as or more numerous than the petals or corolla-lobes** **22**

22a **(from 21b) Stamens as many as and alternate with the 5-10 corolla-lobes; ovary with 5-10 carpels but twice as many locules, each locule with a single ovule; plants parasitic** .... **Lennoa Family, LENNOACEAE**

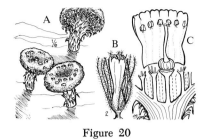

Figure 20

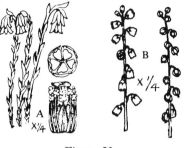

Figure 21

Figure 20 *Ammobroma sonorae* Torr., Sand Bread; A, habit; B, external view of flower; C, dissected flower, in place. This species was once a fairly important item of diet for Indians in southern California.

There are only five species of Lennoaceae, forming a distinctive family with three genera (*Lennoa, Ammobroma* and *Pholisma*) native to northern Mexico and southwestern United States. They are thought to be related to the Boraginaceae and Hydrophyllaceae.

22b Stamens twice as many as the (3) 4-5 (6) petals, these distinct or connate into a lobed tube; ovary with (4) 5 (6) carpels and as many locules (or intruded, partial partitions); ovules numerous; plants mycotrophic, living symbiotically with a fungus; anthers opening lengthwise, usually by longitudinal slits (occasional leafless plants of *Pyrola,* in the related family Pyrolaceae, would be sought here except that the anthers open by seemingly terminal pores) .... Indian Pipe Family, MONOTROPACEAE

Figure 21 A, *Monotropa uniflora* L., Indian Pipe; B, *Pterospora andromedea* Nutt., Pine Drops.

Members of the Monotropaceae get their food from a mycorhizal fungus that forms a living bridge into the roots of forest trees. They are indirectly parasitic on the trees, through the medium of the fungus.

There are about 45 species of Monotropaceae, most of them in temperate or cool parts of the Northern Hemisphere. The Monotropaceae and Pyrolaceae are closely related to the much larger family Ericaceae (see fig. 204), and many botanists treat these two smaller groups as subfamilies of the Ericaceae.

23a (from 16b) Flowers with two or more distinct pistils 24

23b Flowers with a single pistil, composed of one or more carpels 46

24a (from 23a) Trees, shrubs, or woody vines 25

24b Herbs or herbaceous vines, seldom half-shrubs, rarely (*Clematis,* in the Ranunculaceae) woody vines, then with perfect flowers 38

25a (from 24a) Woody vines; flowers small, unisexual 26

25b Erect shrubs or trees; flowers large or small, perfect or unisexual 28

26a (from 25a) Leaves compound; ovules more or less numerous; stamens and staminodes each six; pistils mostly three (–5); fruit fleshy, without a bony inner layer; seeds with small, straight embryo and copious endosperm ...... Lardizabala Family, LARDIZABALACEAE

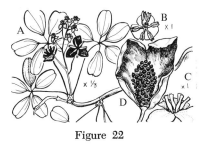

Figure 22

Figure 22 *Akebia quinata* (Houtt.) Decne. A, habit, with pistillate flowers at the bottom of the raceme, and staminate ones toward the top; B, staminate flower; C, pistillate flower, with several separate pistils; D, opened ripe fruit, with numerous seeds.

This plant is grown as a porch vine in eastern United States, where it occasionally escapes from cultivation. It provides good laboratory material to demonstrate an unsealed, merely folded carpel, considered to be a primitive type. The berries are edible.

There are about 30 species of Lardizabalaceae, native to eastern Asia and to Chile. The family is considered to be related to the Menispermaceae (see fig. 24).

26b Leaves mostly simple 27

27a (from 26b) Pistils mostly 12 or more; ovules 2-several; stamens 4-80; fruit fleshy, without a bony layer; seeds with small, straight embryo and copious endosperm; leaves pinnately veined .... Schisandra Family, SCHISANDRACEAE

Figure 23

Figure 23 *Schisandra glabra* (Brickell) Rehder, American Schisandra, is a rare vine of southeastern United States. Each pistillate flower produces several pea-sized red berries on a somewhat elongate receptacle.

Most of the 50 species of this family grow in eastern Asia.

27b Pistils mostly three or six; ovules two, one soon abortive; stamens mostly six, sometimes more numerous; fruit often fleshy, but always with a bony inner layer, commonly more or less falcate-curved; seeds often horseshoe-shaped, commonly with a large, curved embryo and little or no endosperm; leaves often palmately veined .... Moonseed Family, MENISPERMACEAE

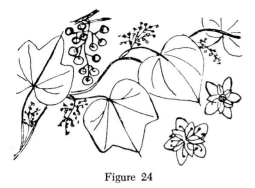

Figure 24

Figure 24 *Menispermum canadense* L., Canada Moonseed, is grape-like in aspect, but has numerous (12-24) stamens and its vine lacks tendrils; the small, fleshy, blue-black fruits contain a single stone.

There are about 400 species of Menispermaceae, most of them in tropical or subtropical regions. *Cocculus carolinus* (L.) DC., the Carolina Moonseed, looks something like *Menispermum canadense*, but has red fruits and only six stamens. Members of the moonseed family are poisonous; some of the tropical species are used to make curare, an arrow poison.

28a (from 25b) Perianth very small and inconspicuous, or none; flowers unisexual; leaves palmately or pinnipalmately veined, with stipules; stamens 3-13 29

28b Perianth more or less well developed, often conspicuous; flowers mostly perfect; stamens generally more or less numerous; leaves of diverse types 30

29a (from 28a) Fruit of numerous small, densely hairy achenes grouped into a tight, globose head; ovule solitary (2); stamens 3-4 (-7); plants monoecious; petiole mitriform at the base, enclosing the

axillary bud .... Sycamore Family, **PLATANACEAE**

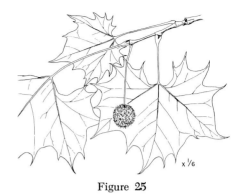

Figure 25

Figure 25 *Platanus occidentalis* L., Sycamore, Plane tree.

The family Platanaceae has only one genus and half a dozen species, all with maple-like but alternate leaves. The large branches of the tree have a pale, blotchy look because the bark peels off in large, irregular pieces. The pistillate flowers have several (commonly 5-8) separate carpels; the upper part of each ovary is unsealed, with the margins of the carpel merely pressed together. The fossil record of the Sycamores can be traced back into the Lower Cretaceous period, more than 100 million years ago.

29b Fruit of several follicles; ovules numerous; stamens 8-13; plants dioecious; petiole not mitriform .... Cercidiphyllum Family, **CERCIDIPHYLLACEAE**

Figure 26

Figure 26 *Cercidiphyllum japonicum* Sieb. & Zucc., a Japanese species occasionally cultivated in the United States. A, twig with staminate flowers, before the leaves develop; B, C, twigs with pistillate flowers, before the leaves develop; D, leafy twig, with opened follicles (fruits).

The family has only one genus, with two species, one Japanese, the other native to China. The leaves are rounded-cordate and resemble those of *Cercis*, the Redbud, whence the name *Cercidiphyllum*.

30a (from 28b) Flowers evidently perigynous, with a well developed hypanthium 31

30b Flowers hypogynous, without a hypanthium 33

31a (from 30a) Leaves generally with well developed stipules, usually alternate, only seldom opposite; seeds without an aril .... Rose Family, ROSACEAE

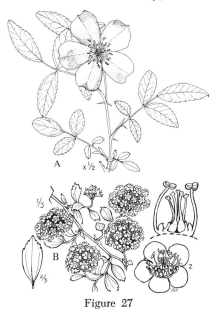

Figure 27

Figure 27 A, *Rosa carolina* L., Wild Rose; B, *Spiraea vanhouttei* (Briot) Zabel, Bridal Wreath, showing habit, separate leaf, an enlarged flower, and a detail of part of a flower, with five separate pistils and two of the many stamens.

The large, highly diversified family Rosaceae has about 3,000 species, found throughout most of the world, but most common in temperate and subtropical parts of the Northern Hemisphere. Some members are woody, some herbaceous; some have simple leaves, some compound; some have separate pistils, some have a single simple pistil, and some have a compound pistil and an inferior ovary. All have perigynous (or less commonly epigynous) flowers, and nearly all have stipulate leaves and numerous stamens. This combination of features is restricted mainly to the Rosaceae, the most noteworthy exception being some of the Mimosaceae, which have compound leaves and flowers wih only a single carpel. Members of the Rosaceae with unicarpellate flowers, such as *Prunus*, have simple leaves. *Kerria*, *Holodiscus* and *Physocarpus* are some of the genera of the Rosaceae, in addition to *Rosa* and *Spiraea*, that will key here. See also figs. 42, 92, 182.

31b Leaves without stipules (except in *Apacheria*, of the Crossosomataceae, which has opposite leaves and arillate seeds) 32

32a (from 31b) Perianth of numerous slender, more or less petaloid tepals, spirally arranged on the outside of a large, vase-shaped, narrow-mouthed hypanthium; fruit of achenes within the enlarged hypanthium; filaments short or none; plants aromatic .... Spice-bush Family, CALYCANTHACEAE

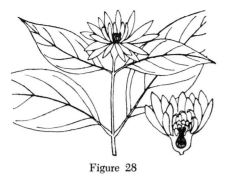

Figure 28

Figure 28 *Calycanthus floridus* L., Spice-bush, or Strawberry Shrub. There are only half a dozen species of Calycanthaceae, confined to China and temperate North America. The leaves have a spicy odor, and the dark-maroon flowers smell something like strawberries.

32b **Perianth biseriate, of four or five sepals and as many petals attached to the margin of a saucer-shaped or cup-shaped, wide open hypanthium; fruit of follicles, the seeds arillate; filaments well developed; plants not aromatic ...... Crossosoma Family, CROSSOSOMATACEAE**

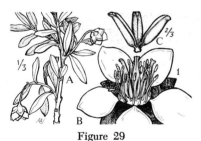

Figure 29

Figure 29 *Crossosoma californicum* Nutt.; A, habit; B, flower; C, three follicles, after opening.

There are only three genera and about ten species of Crossosomataceae, all xerophytic shrubs of the southwestern United States and adjacent Mexico. *Forsellesia* (*Glossopetalon*)

has often been included in the family Cela-straceae, but it has separate carpels (one or more) and has recently been appropriately transferred to the Crossosomataceae. The third genus, *Apacheria*, was discovered only a few years ago and named in 1975. It has opposite, stipulate leaves, in contrast to the alternate, exstipulate leaves of *Crossosoma*. The leaves in all three genera are small.

33a **(from 30b) Stamens 5-10; ovaries mostly five, each with two ovules and a deeply lateral, almost basal style; maritime shrub .... Bay Cedar Family, SURIANACEAE**

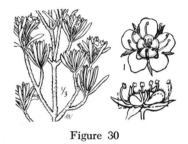

Figure 30

Figure 30 *Suriana maritima* L., Bay Cedar, grows along coastal beaches in southern Florida and along tropical coasts in general.

The family has only four genera and six species. Aside from *Suriana maritima*, the species are all Australian.

33b **Stamens usually more or less numerous (more than 10); ovaries often more than five; ovules 1-many; styles terminal or wanting; shrubs or trees; not maritime 34**

34a **(from 33b) Leaves with stipules; receptacle conspicuously elongate .... Magnolia Family, MAGNOLIACEAE**

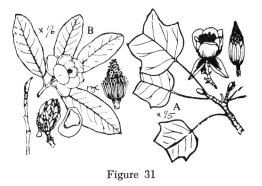

Figure 31

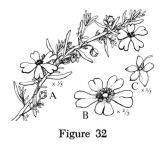

Figure 32

Figure 31 A, *Liriodendron tulipifera* L., Tulip Tree; B, *Magnolia virginiana* L., Sweet Bay.

There are about twelve genera and about 200 species of Magnoliaceae, most of them in warm regions, but a few in temperate climates. *Magnolia* has unusual fruits; the initially separate carpels become coherent on the enlarged receptacle, and each carpel opens to the outside. The Tulip Tree has distinctive leaves that are broadly notched across the summit. There are only two species of *Liriodendron*, *L. tulipifera* in eastern United States, and *L. chinense* (Hemsley) Sarg. in China. This pair of species illustrates the well known floristic relationship between the eastern United States (especially the southern Appalachian region) and eastern Asia.

34b Leaves without stipules; receptacle small, not elongate 35

35a (from 34b) Stamens with well differentiated filament and anther; plants without ethereal oil cells, not aromatic; seeds mostly arillate .... Dillenia Family, DILLENIACEAE

Figure 32 *Hibbertia hypericoides* Benth., a shrub of western Australia; A, habit; B, flower, from above; C, calyx.

There are about 350 species of Dilleniaceae. Most of them are tropical or subtropical, and they are especially abundant in Australia. Flowers of the Dilleniaceae are much like those of Magnoliaceae and related families, but the plants do not have the characteristic ethereal oil cells of the latter group.

35b Stamens with very short, stout filament, or scarcely divided into filament and anther; plants aromatic, with scattered ethereal oil cells in the softer tissues .. 36

36a (from 35b) Perianth mostly trimerous; endosperm ruminate; seeds often arillate .... Custard-Apple Family, ANNONACEAE

Figure 33

Figure 33 *Asimina triloba* (L.) Dunal, Paw-paw, native to much of the eastern United States, has large, fleshy, edible fruits up to 15 cm long.

There are about 2300 species of Annonaceae; except for *Asimina* the family is almost entirely tropical. Members of the Annonaceae have extraordinarily diversified pollen, but none of it has the three apertures found in most families of dicotyledons.

36b Perianth not trimerous; endosperm not ruminate; seeds not arillate 37

37a (from 36b) Wood without vessels; pollen grains in tetrads, each grain with a single furrow; ovules 1-several; fruit variously fleshy or dry and dehiscent, but not of 1-seeded follicles .... Drimys Family, WINTERACEAE

Figure 34

Figure 34 *Drimys winteri* J. R. & G. Forst., a shrub native to Andean region of South America.

There are about 100 species of Winteraceae, all tropical or of the Southern Hemisphere, most of them on the Pacific Islands. The Winteraceae are generally regarded as the most archaic existing family of flowering plants, the least modified from ancestral early Cretaceous angiosperms.

37b Wood with vessels; pollen grains in monads, each with three furrows; ovule solitary; fruit dry, consisting of 1-seeded follicles .... Illicium Family, ILLICIACEAE

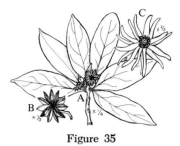

Figure 35

Figure 35 *Illicium floridanum* Ellis, Florida Star-Anise, a shrub native to southeastern United States. A, flowering twig; B, cluster of opened follicles, from above; C, flower, from above.

There are about forty species of Illiciaceae, all in the genus *Illicium.* Most of the species occur in southeastern Asia or in the Caribbean region. Some are traditional sources of anethole (anise oil), which is also obtained from *Pimpinella anisum* L., of the very different family Apiaceae.

38a (from 24b) Plants aquatic, rooted in the substrate, with long-petiolate leaves and peltate blades floating or raised above the surface of the water (some dissected submerged leaves also present in *Cabomba*); flowers solitary from the axils, long-pedunculate, borne at or a little above the surface of the water 39

38b Plants terrestrial or sometimes aquatic, but if aquatic then without peltate leaf-blades; flowers various, but not long-pedunculate from the axils of a submerged stem 40

39a (from 38a) Flowers large and with numerous parts, the sepals and petals collectively about 20-30, the stamens about 200 or more; pistils individually embedded in the enlarged, obconic receptacle; ovule 1 (2); seeds with large embryo, no perisperm, and virtually no endosperm .... Lotus Lily Family, NELUMBONACEAE

Figure 36

Figure 36 *Nelumbo lutea* (Willd.) Pers., American Lotus Lily.

There are only three species of Nelumbonaceae, all in the genus *Nelumbo. Nelumbo* is famous for its very long-lived seeds, which can retain their vitality for 1500 years or possibly much more.

39b Flowers rather small and with fewer parts, the sepals and petals collectively (4) 6 (-8), the stamens rarely more than 17; pistils borne on the surface of the small receptacle, not embedded; ovules (1) 2-3; seeds with small embryo, copious perisperm, and scanty endosperm .... Cabomba Family, CABOMBACEAE

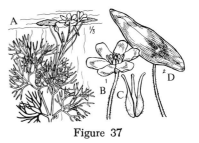

Figure 37

Figure 37 *Cabomba caroliniana* A. Gray, Fanwort; A, habit; B, flower; C, three separate pistils of a flower; D, floating leaf. This species is native to the eastern United States, especially southward.

There are two genera and eight species of Cabombaceae, native chiefly to tropical and warm-temperate regions. *Cabomba* has cream-colored, trimerous flowers, with six petaloid tepals (in two sets of three), six stamens, and mostly three pistils; the principal leaves are submersed and dissected, and the floating leaves are small. *Brasenia* has a single, nearly cosmopolitan species, *B. schreberi* Gmel. It has purplish flowers, trimerous as in *Cabomba,* but with 12-18 stamens and 4-8 pistils; the leaves have a floating, peltate blade.

40a (from 38b) Flowers without a perianth, borne in dense, bracteate spikes or racemes; stamens 3-8 .... Lizard's Tail Family, SAURURACEAE

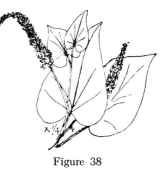

Figure 38

Figure 38 *Saururus cernuus* L., Lizard's Tail. There are five genera and only seven species of Saururaceae, native to North America and eastern Asia. Members of this family have three to five carpels and either three stamens in one cycle or six or eight stamens in two cycles. *Saururus* differs from the other genera in having the carpels virtually distinct. See also fig. 191.

40b Flowers with a more or less well developed perianth, borne in various sorts of inflorescences 41

41a (from 40b) Flowers hypogynous or very nearly so, and without a prominent disk 42

41b Flowers distinctly perigynous, or with a well developed perigynous disk 44

42a (from 41a) Flowers small, unisexual; vines .... Moonseed Family, MENISPERMACEAE
A few species of Menispermaceae are more or less herbaceous and would key here. See fig. 24.

42b Flowers small or large, perfect or seldom unisexual, the viny species always with perfect (and often large) flowers 43

43a (from 42b) Stamens numerous; sepals often quickly deciduous; plants not notably succulent .... Buttercup Family, RANUNCULACEAE

Figure 39

Figure 39 A, *Delphinium tricorne* Michx., Dwarf Larkspur; B, *Ranunculus acris* L., Meadow Buttercup.

There are about 2000 species of Ranunculaceae, most of them native to temperate or boreal regions. Most herbaceous plants with hypogynous flowers, numerous stamens, separate carpels, and exstipulate leaves belong to the Ranunculaceae. The family provides many garden ornamentals, including species of *Adonis*, *Anemone*, *Aquilegia* (Columbine), *Clematis*, Eranthis, *Delphinium* (Larkspur), *Helleborus* (Christmas Rose), and *Ranunculus* (Buttercup). See also fig. 96.

43b Stamens usually as many or twice as many as the petals; sepals persistent; plants distinctly succulent ...... Stonecrop Family, CRASSULACEAE

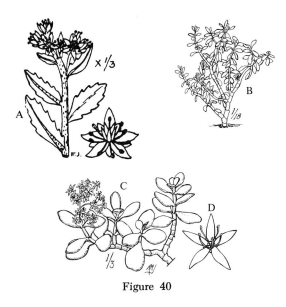

Figure 40

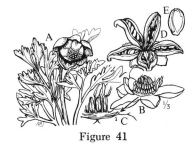

Figure 41

44a (from 41b) Flowers perigynous, with a definite hypanthium 45

44b Flowers with a perigynous disk, but without a hypanthium; stamens numerous; leaves without stipules ....
.... Peony Family, PAEONIACEAE

Figure 40   A, *Sedum telephium* L., Live-for-ever; B-D, *Crassula argentea* Thumb., Jade Plant, B, habit; C, flowering branch; D, flower.

There are about a thousand species of Crassulaceae, most of them in dry parts of the world. Most succulent plants with regular flowers and separate carpels belong to the Crassulaceae. Many species of *Crassula, Kalanchoe, Sedum, Sempervivum,* and other genera are cultivated as garden ornamentals or as house plants. *Kalanchoe* has the petals united toward the base to form a sympetalous corolla, but the other genera mentioned here have separate petals.

Figure 41   *Paeonia brownii* Dougl., Western Peony. A, habit; B, C, portion of flower, showing stamens on a perigynous disk; D, ripe, opened fruits; E, seed.

Paeonia was formerly included in the Ranunculaceae, but botanists now agree that it should form a distinct family allied to the Dilleniaceae. There are about thirty species, most of them in temperate Eurasia, two in western United States. The cultivated Peonies have highly double flowers, with numerous petals; most of them are garden hybrids.

45a (from 44a) Leaves with stipules; stamens numerous .... Rose Family, ROSACEAE

Figure 42

Figure 42 A-F, *Potentilla gracilis* Dougl., Five Fingers. A, B, habit; C, a single pistil of a flower; D, E, dissection of flower parts, E showing perigynous stamens; F, single leaf. G-L, *Agrimonia gryposepala* Wallr., Agrimony. G, flower, with stamens removed; H, detail of central part of flower, showing position of stamens; I, spiny fruit; J, leaf; K, portion of inflorescence; L, habit.

As noted under fig. 27, the Rose family is highly diversified; it must be keyed out in several places. Strawberries (*Fragaria*), Raspberries (*Rubus*, section *Ideobatus*) and Blackberries (*Rubus*, section *Eubatus*) are among the members of the group that will key out here. See figs. 27, 92, 182.

45b Leaves without stipules; stamens 10 or fewer .... Saxifrage Family, **SAXIFRAGACEAE**

Figure 43

Figure 43 *Saxifraga marshallii* Greene, a Pacific Coast species; A, habit; B, portion of glandular-hairy stem; C, leaf; D, flower; E, stamen, with expanded filament; F, mature flower, with persistent, reflexed sepals and two opened follicles. *S. marshallii* has the two carpels separate virtually to the base, but most species of Saxifragaceae have the carpels connate at least in the lower part, and would key elsewhere. See figs. 80, 109, 153.

46a (from 23b) Petals distinct from each other, or wanting **47**

46b Petals united, at least toward the base, to form a sympetalous corolla **204**

47a (from 46a) Plants distinctly aquatic, either wholly submerged, or partly floating, or with less than half of the length of the stem elevated above the water .. **48**

47b Plants terrestrial, or sometimes growing in shallow water, but then more than half of the stem elevated above the water **57**

48a (from 47a) Plants of rapidly flowing water, more or less thalloid, the parts not easily recognizable as roots, stems, and leaves; ovules numerous; fruit capsular; seeds numerous and tiny .... River Weed Family, **PODOSTEMACEAE**

Figure 44

Figure 44 *Podostemum ceratophyllum* Michx., River Weed.

There are about 200 species of Podostemaceae, most of them in tropical countries, only a few in the United States. The plants grow submersed or with some parts floating, and produce aerial flowers and fruits at times of low water. The Podostemaceae have no close allies. Botanists believe that they are distantly related to the order Rosales.

48b Plants of quiet water, not thalloid; other features various **49**

49a (from 48b) Plants with large, solitary, showy flowers and floating, peltate to basally cordate or hastate leaf-blades; stamens numerous; ovary with 5-35 carpels and locules .... Water Lily Family, **NYMPHAEACEAE**

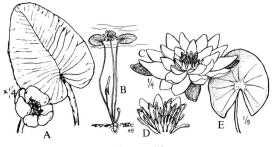

Figure 45

Figure 45   A, *Nuphar advena* Aiton, Yellow Water Lily; B-D, *Nymphaea odorata* Aiton, Water Lily. B, habit; C, flower; D, detail of portion of flower, showing stigmas (front), stamens, and petals; E, leaf.

There are about fifty species of Nymphaeaceae, found throughout most of the world. The tropical Royal Water Lily (*Victoria*) is another well known genus of the family. The water lilies belong to an ancient lineage that probably originated late in the Lower Cretaceous period. They resemble the monocotyledons in some respects, and many botanists consider that they stand close to the dividing line between monocots and dicots.

49b Plants with small, often inconspicuous flowers and without floating leaves that are peltate or basally cordate or has-

tate; stamens not more than 10 (12), except usually in Ceratophyllaceae; ovary with not more than five carpels and locules **50**

50a (from 49b) Pistil with a single style and stigma **51**

50b Pistil with two or more stigmas, and often with two or more styles **54**

51a (from 50a) Plants free-floating **52**

51b Plants rooted in the substrate, usually more or less emergent; leaves narrow, simple and entire **53**

52a (from 51a) Principal leaves with spongy-inflated, floating petiole and rhombic, toothed, aerial blade; flowers with four sepals, four petals, and a bilocular ovary with one ovule in each locule (only one ovule maturing in the fruit); stamens four; fruit large (commonly chestnut-sized) and with two or four prominent horns or spines .... Water Chestnut Family, **TRAPACEAE**

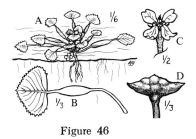

Figure 46

Figure 46 *Trapa natans* L., Water Chestnut, is used in specialty dishes, particularly in Chinese cookery. A, floating plant; B, leaf, with inflated petiole; C, flower; D, fruit.

The family has only the genus *Trapa,* with about fifteen species, all native to the Old World.

52b Leaves all submersed, sessile, dissected; flowers without petals; ovary unilocular, with a single ovule; stamens (5) 10-20 (27); fruit small, unarmed .... Hornwort Family, CERATOPHYLLACEAE

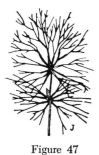

Figure 47

Figure 47 *Ceratophyllum demersum* L., Hornwort, is often used in aquaria to provide aeration.

The family has only a single genus and half a dozen species, but they occur in nearly all parts of the world. The Ceratophyllaceae are thought to be florally reduced relatives of the Cabombaceae, which have somewhat similar submersed, dissected leaves.

53a (from 51b) Leaves whorled; stamen one; ovary unilocular, with a single ovule .... Mare's Tail Family, HIPPURIDACEAE

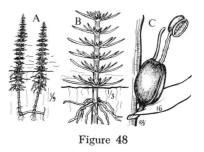

Figure 48

Figure 48 *Hippuris vulgaris* L., Mare's Tail, is the only species of its family. A, habit; B, portion of plant; C, flower, with naked ovary, single style, and epigynous stamen. Mare's Tail is widely distributed in temperate and boreal regions of the Northern Hemisphere, and in Australia and southern South America. On chemical and embryological grounds the Hippuridaceae belong with the subclass Asteridae, in which most families have a sympetalous corolla.

53b Leaves opposite; stamens four; ovary bilocular, with several ovules in each locule .... Loosestrife Family, LYTHRACEAE

Figure 49

Figure 49 *Didiplis diandra* (Nutt.) A. Wood, Water Purslane, is unusual in its family in being a submerged aquatic. Most other members of the Lythraceae are terrestrial, or strongly emergent from shallow water, and will key elsewhere. In spite of its name, this species normally has four stamens, like some other members of its family. See figs. 100, 108, 151.

54a **(from 50b) Leaves with well developed stipules that form a thin sheath (ochrea) around the stem at the base of the petiole; ovule solitary in the single locule of the ovary; leaves simple, alternate; stigmas three** .... **Buckwheat Family, POLYGONACEAE**

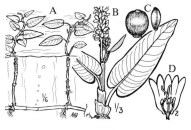

Figure 50

Figure 50 *Polygonum amphibium* L., Water Smartweed, is one of several amphibious species of its family. A, habit, with floating stem-tips; B, upper part of plant, with leaves, sheathing stipules, and inflorescence; C, achene in two views; D, detail of flower, with perianth laid open. Water Smartweed can grow on dry land, or in water several feet deep, or often in places where the water level changes during the season. Most species of Polygonaceae are terrestrial and will key elsewhere. See figs. 89, 148.

54b **Leaves with small, inconspicuous stipules that do not form a sheath, or often without stipules; ovary with two or more locules, each with one or more ovules** **55**

55a **(from 54b) Ovules several or many in each of the 2-5 chambers of the ovary; flowers perfect, and with both sepals and petals; leaves opposite or whorled, entire or merely toothed** .... **.... Waterwort Family, ELATINACEAE**

Figure 51

Figure 51 *Elatine triandra* Schkuhr, Waterwort, grows in shallow water in much of the North Temperate Zone.

There are about forty species of Elatinaceae, but they are inconspicuous and easily overlooked. The family has only two genera. The flowers of *Bergia* are pentamerous, whereas those of species of *Elatine* are dimerous, trimerous, or tetramerous.

55b **Ovules solitary in each of the 2-4 chambers of the ovary; flowers perfect, or more often unisexual** **56**

56a **(from 55b) Perianth (at least the sepals) present; stamens three, four, or eight; leaves entire to dissected, variously alternate, opposite, or whorled** .... **Water Milfoil Family, HALORAGACEAE**

Figure 52

Figure 52 *Myriophyllum spicatum* L., Spiked Water Milfoil, is the commonest species of *Myriophyllum* in the northern United States; it is circumboreal, growing also in Europe, northern Asia, and Canada.

There are about 125 species of Haloragaceae, widely distributed throughout the world. *Myriophyllum brasiliense* Cambess, sometimes called Parrot's Feather, is a well known aquarium plant.

56b Perianth wanting; stamen one, or seldom stamens 2-3; leaves entire, opposite or seldom whorled .... Water Starwort **Family, CALLITRICHACEAE**

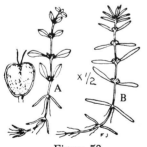

Figure 53

Figure 53 *Callitriche palustris* L., Water Starwort.

The family has only the genus *Callitriche,* with about thirty-five species, found in most parts of the world. Although most species are aquatic, a few grow in mud or on very wet soil. Like the Hippuridaceae, the apetalous Callitrichaceae are referred to the mostly sympetalous subclass Asteridae on chemical and embryological grounds.

57a **(from 47b) Flowers borne in small inflorescences that closely resemble an individual flower, each such inflorescence (called a cyathium) composed of a cup-shaped involucre with 1-5 evident glands around the margin (the gland sometimes with a spreading, petal-like appendage); from each cyathium protrude several or many stamens and one stipitate, 3-lobed and 3-locular ovary with three or more styles; plants with a poisonous, irritating milky juice (*Euphorbia*) ....**
.... Spurge Family, EUPHORBIACEAE

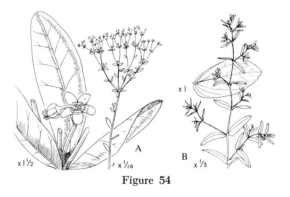

Figure 54

Figure 54   A, *Euphorbia corollata* L., Flowering Spurge, is a common and widespread species in eastern United States; B, *Euphorbia maculata* L., Wartweed, is an eastern American species that has now become an almost cosmopolitan weed.

The genus *Euphorbia,* with about 1500 species, represents the culmination of a trend within its family toward floral reduction and aggregation to produce inflorescences that resemble single flowers. Although species of *Euphorbia* are relatively stereotyped in floral structure, they are extraordinarily diverse in habit; some botanists would divide the genus into several genera. *Poinsettia,* familiar as a Christmas-time decorative plant, is one of these segregate genera. The most bizzare species are succulent, leafless, spiny African plants that vegetatively resemble some of the larger cacti. A quick thumbnail test: if it has milky juice it is a euphorbiad and not a cactus. See also fig. 69.

57b   Flowers not as above; juice watery or less often milky **58**

58a   (from 57b) Flowers individually very small, with much reduced or no perianth, some or all of them borne in compact, spike-like clusters or in dense heads; nearly all woody plants, except many of the Piperaceae and a few of the Moraceae; flowers unisexual, except in many Piperaceae **59**

58b   Flowers otherwise, either perfect or unisexual, with or without a well developed perianth, but if unisexual and with reduced or no perianth, then not borne in compact, spike-like clusters or dense heads; habit various **71**

59a   (from 58a) Leaves pinnately compound **60**

59b   Leaves simple, variously entire or toothed to deeply lobed   **61**

60a   (from 59a) Leaves alternate; plants aromatic, resinous; fruit a nut or a dry drupe .... **Walnut Family, JUGLANDACEAE**

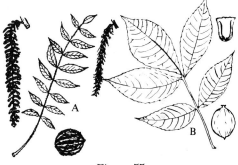

Figure 55

Figure 55   A, *Juglans nigra* L., Black Walnut; B, *Carya ovata* (Miller) K. Koch, Shagbark Hickory.

There are seven or eight genera of Juglandaceae and about sixty species, mostly in temperate and subtropical parts of the Northern Hemisphere. The family can be recognized by the combination of woody habit, alternate, compound leaves, and tiny, unisexual flowers, the staminate ones in catkins. The Juglandaceae are important for their hard, durable wood and edible nuts. Species of Hickory (*Carya*) are common components of eastern American forests.

60b   Leaves opposite; plants neither aromatic nor resinous; fruit a samara .... **Olive Family, OLEACEAE**

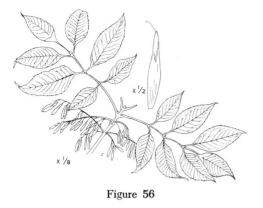

Figure 56

Figure 56 *Fraxinus americana* L., White Ash, is a common and valuable forest tree in eastern United States. Most members of the Oleaceae have an evident, sympetalous corolla and will key elsewhere. See fig. 228.

61a (from 59b) Plants with milky juice; leaves with stipules, and often with cystoliths in the epidermis ....
.... **Mulberry Family, MORACEAE**

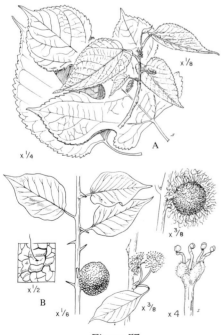

Figure 57

Figure 57 A, *Morus rubra* L., Red Mulberry; B, *Maclura pomifera* (Raf.) Schneid., Osage Orange. Mulberry fruits are eaten by birds and people, and the leaves are eaten by silkworms.

There are about 1,500 species of Moraceae, most of them tropical or subtropical. Two-thirds of the species belong to the Fig genus, *Ficus. Ficus elastica* Roxb., Rubber Plant, is widely grown in home and greenhouse. Visitors to the American tropics may see *Cecropia*, a tree with large leaves that are palmately very deeply lobed, at first appearing to be compound. *Artocarpus*, the Breadfruit, is another familiar genus. The Moraceae can be recognized by their usually woody habit, milky juice, and tiny, unisexual, apetalous flowers, the pistillate flowers with a single ovary and two styles or two style branches.

61b Plants without milky juice; leaves with or without stipules, but without cystoliths **62**

62a (from 61b) Leaves opposite or whorled **63**

62b Leaves alternate **65**

63a (from 62a) Leaves whorled, much reduced and scale-like .... **Australian Pine Family, CASUARINACEAE**

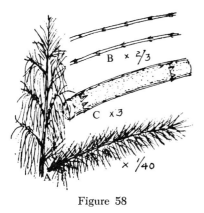

Figure 58

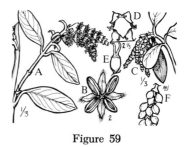

Figure 59

Figure 58 *Casuarina equisetifolia* L., Australian Pine; A, branches with long, slender, green branchlets; B and C, portions of branchlets, showing scale-like leaves.

The family has only the genus *Casuarina,* with about fifty species, most of them native to Australia; some are cultivated in southern United States and in Mexico as street trees. The branchlets recall stems of *Equisetum* (Horsetail), but are anatomically very different. In Australia they are called She oak.

63b   **Leaves opposite, of ordinary (though sometimes rather small) size, not scale-like** **64**

64a   **(from 63b) Leaves leathery; ovary unilocular, with two ovules pendulous from the wall of the upper part of the locule; seeds with very small embryo and copious endosperm; plants not maritime** ...... **Silk Tassel Family, GARRYACEAE**

Figure 59 *Garrya fremontii* Torr., Silk Tassel. A, habit of male plant; B, staminate flower; C, habit of female plant; D, portion of drooping pistillate catkin; E, pistillate flower, with naked ovary and two styles; F, portion of drooping fruiting inflorescence.

The family has only the genus *Garrya,* with about fifteen species, native mainly to western United States. Botanists now believe that *Garrya* is related to the Dogwoods (Cornaceae), differing mainly in its strongly reduced flowers.

64b   **Leaves succulent, rather small; ovary 4-locular, with a single parietal-basal ovule in each locule; embryo filling the seed; endosperm wanting; plants maritime** .... **Batis Family, BATACEAE**

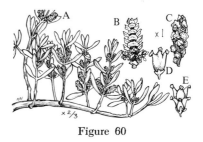

Figure 60

Figure 60 *Batis maritima* L. A, habit; B, staminate inflorescence; C, pistillate inflorescence; D, pistillate flower; E, developing fruits.

The family has only the genus *Batis,* with two species, native to tropical and subtropical coasts. The taxonomic affinities of the Bataceae are in dispute. The fact that they produce mustard oil provides a possible clue to their relationships.

65a **(from 62b) Fruits capsular, dehiscent; seeds small; ovules few to often numerous** **66**

65b **Fruits indehiscent, retaining the few or solitary, rather large seeds; ovules few or solitary** **67**

66a **(from 65a) Seeds numerous, plumehairy; ovary and capsule unilocular** .... **Willow Family, SALICACEAE**

Figure 61

Figure 61 A, *Populus tremuloides* Michx., Quaking Aspen; B, *Salix exigua* Nutt., Sandbar Willow.

The family Salicaceae has only two genera and fewer than 400 species, but some of them are very common, especially in moist places in North Temperate regions. The family can be recognized by its woody habit, unisexual flowers borne in catkins, and numerous ovules and seeds. Many poplars and willows are cultivated as street and lawn trees; they grow

rapidly but have soft wood and are short-lived. Some arctic and alpine willows form mats on the ground, instead of growing upright.

66b **Seeds few, not plumed** .... **Witch Hazel Family, HAMAMELIDACEAE**

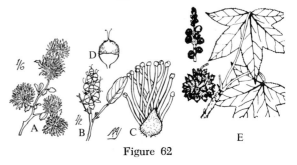

Figure 62

Figure 62 A-D, *Fothergilla gardenii* Murray, Witch Alder, is a low shrub of the Atlantic Coastal Plain. A, staminate inflorescences; B, pistillate inflorescence; C, staminate flower, with subtending bract; D, pistillate flower. E, *Liquidambar styraciflua* L., Sweet Gum, is a common tree in wet woods in much of the eastern United States. Some other members of the small family Hamamelidaceae, such as *Hamamelis,* have definite petals and will key elsewhere.
See fig. 197.

67a **(from 65b) Ovary unilocular, and with a single ovule** **68**

67b **Ovary with 2-several locules, at least toward the base (often unilocular toward the summit); ovules 2-several, one in each locule** **70**

68a **(from 67a) Seed with a minute, scarcely visible embryo; mostly herbs and halfshrubs, seldom more distinctly woody; flowers in fleshy, spadix-like spikes** .... **Pepper Family, PIPERACEAE**

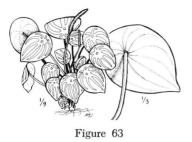

Figure 63

Figure 63 *Peperomia sandersii* C. DC. is one of several species of *Peperomia* frequently grown indoors for its ornamental foliage.

There are about 1500 species of Piperaceae, all tropical. Only two genera, *Piper and Peperomia*, make up the bulk of the family. The ground-up fruits of *Piper nigrum* L. are the source of pepper.

68b Seed with a well developed embryo; woody plants; flowers in catkins, the inflorescence not fleshy (except as noted under lead 69b) **69**

69a (from 68b) Ovary with a single style and stigma; ovule pendulous from near the summit of the ovary; plants not aromatic .... **Corkwood Family, LEITNERIACEAE**

Figure 64

Figure 64 *Leitneria floridana* Chapm, Corkwood.

This is the only species of Leitneriaceae. A native of the southeastern United States, it is important mainly as a plague to botanists, who are uncertain about its relationships.

69b Ovary with two styles or two evident style-branches; ovule basal and erect; plants aromatic, with resin-dotted leaves; seed filled by the straight embryo, nearly or quite without endosperm and perisperm. (A few members of the Chenopodiaceae, such as *Sarcobatus vermiculatus*, might be sought here except for their non-aromatic nature and for having a spirally coiled or circular and peripheral embryo; these species grow in saline habitats and have somewhat fleshy inflorescence-spikes; see fig. 128) .... **Bayberry Family, MYRICACEAE**

Figure 65

Figure 65 *Myrica gale* L. Sweet Gale.

There are three genera (*Myrica, Canacomyrica,* and *Comptonia*) and about fifty-five species of Myricaceae, widespread in temperate and subtropical regions. Like the legumes, they commonly harbor nitrogen-fixing bacteria in their roots.

70a (from 67b) Female flowers (as well as the male) borne in catkins; fruits without a cupule, or sometimes (e.g., in *Corylus*) with a foliaceous hull derived from two or three bracts; carpels two .... **Birch Family, BETULACEAE**

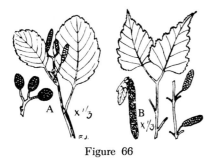

Figure 66

Figure 66   A, *Alnus tenuifolia* Nutt., Thin-leaved Alder; B, *Betula populifolia* Marsh., Gray Birch.

The family Betulaceae has about 130 species, in six genera: *Alnus* (Alder), *Betula* (Birch), *Carpinus* (Hòrnbeam), *Corylus* (Hazelnut), *Ostrya* (Hop Hornbeam), and *Ostryopsis*. Most of the species grow in temperate and cold parts of the Northern Hemisphere. In Canada and northern United States, birches such as *Betula papyrifera*, (Paper Birch) and *B. populifolia* are mainly seral trees, coming in and growing rapidly after fire or other disturbance.

70b   **Female flowers not in catkins, generally subtended or enclosed individually or in small groups by a characteristic hull or cupule that often appears to represent numerous reduced bracts; carpels three or sometimes six .... Beech Family, FAGACEAE**

Figure 67

Figure 67   A, *Quercus alba* L., White Oak; B, *Fagus grandifolia* Ehrh., American Beech.

Oaks are important constituents of American forests. They provide strong and beautiful wood. Most oaks have pinnately lobed leaves, but in some species the leaves are merely toothed, or even entire; the characteristic acorn, with a basal cup, provides a ready marker for the genus. Beech is easily recognized in the forest by its smooth, close, gray bark and merely toothed leaves. The American Chestnut (*Castanea dentata* (Marshall) Borkh.), formerly a valuable forest tree, has been nearly wiped out by chestnut blight.

71a   **(from 58b) Ovules 2-5, pendulous from the top of a free-central placenta, one cupped at the base in each of the semi-locules of the ovary, which has partitions at the base only; woody plants .... Olax Family, OLACACEAE**

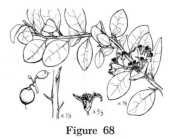

Figure 68

Figure 68   *Ximenia americana* L., American Ximenia, is a tropical American species with valuable wood, edible fruits, and seeds that are rich in oil.

There are about 250 species of Olacaceae, all of tropical or subtropical regions. Many or most of them are hemi-parasites, green-leaved and rooted in the ground, but attached to the roots of other plants.

71b Ovules and ovary not as above in all respects; plants woody or herbaceous 72

72a (from 71b) Ovary superior (or nude) .. 73

72b Ovary inferior, or at least half-inferior 180

73a (from 72a) Flowers regular or nearly so 74

73b Flowers evidently irregular 165

74a (from 73a) Flowers unisexual and with much-reduced or no perianth; ovary with three locules, each containing one or two apical-axile, pendulous ovules 75

74b Flowers differing in one or another respect from those of the previous group 77

75a (from 74a) Ovules epitropous, the raphe ventral (next to the axis); ovary topped by a central style, or the styles or stigmas clustered on the center of the top of the ovary; plants very often with milky juice .... Spurge Family, **EUPHORBIACEAE**

Figure 69

Figure 69 A, *Croton glandulosus* L., Glandular Croton, is a weedy, mostly tropical and subtropical species that also grows in much of the southeastern United States; B-F, *Ricinus communis* L., Castor Bean, is a coarse plant often cultivated for its large, ornamental leaves; its seeds are highly poisonous and are the source of castor oil. B, habit; C, pistillate flower; D, staminate flower; E, fruits; F, seed.

Manihot (Cassava) and *Hevea* (Para Rubber) are some other well known members of this large family, which has about 7,500 species. Anything with milky juice, unisexual flowers, and a trilocular ovary will very probably belong to the Euphorbiaceae. See also fig. 54

75b Ovule apotropous, the raphe dorsal (on the side away from the axis); ovary capped by distinct, marginal or submarginal styles; juice not milky 76

76a (from 75b) Ovules two per locule; seeds generally with copious endosperm; mesophytes .... Box Family, **BUXACEAE**

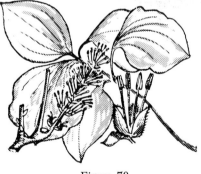

Figure 70

Figure 70 *Pachysandra procumbens* Michx., Alleghany Spurge, grows in rich woods in southeastern United States. Another species of *Pachysandra, P. terminalis* Sieb. & Zucc., Japanese Pachysandra, is often planted as a ground-cover for shady spots.

Buxus sempervirens L., Boxwood, is another member of this small (sixty species) but cosmopolitan family.

76b Ovules solitary in each locule; seeds with scanty or no endosperm; desert shrubs .... Jojoba Family, SIMMONDSIACEAE

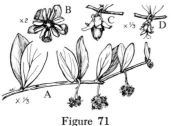

Figure 71

Figure 71 *Simmondsia chinensis* (Link) C. K. Schneid., Jojoba Bush. A, habit of male plant; B, staminate flower; C, fruit; D, pistillate flower. The species is a desert shrub of southwestern United States and adjacent Mexico (not China, in spite of the name!), and is the only member of its family. Jojoba seeds contain a valuable liquid wax. *Simmondsia* has often been included in the Buxaceae, but more recent students consider that the relationship is probably rather remote.

77a (from 74b) Leaves with scattered, embedded secretory cavities that appear as translucent dots when the leaf is held up to the light 78

77b Leaves without such cavities, not appearing translucent-dotted 79

78a (from 77a) Leaves simple and usually entire, always opposite or whorled; ovules two to more often several or numerous in each locule of the unilocular or more often plurilocular ovary; flowers without an evident nectary disk or a gynophore .... St. John's-Wort Family, CLUSIACEAE

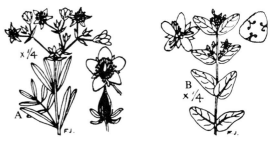

Figure 72

Figure 72 A, *Hypericum prolificum* L., Shrubby St. John's-Wort, is common in eastern United States; B, *Triadenum virginicum* (L.) Raf., Marsh St. John's-Wort, is another common eastern species.

There are about 900 species of Clusiaceae, some in the United States but most of them in moist, tropical countries. The group includes herbs, shrubs, and trees. Anything with simple, opposite, glandular-punctuate leaves and hypogynous flowers with numerous stamens and a compound pistil very probably belongs to the Clusiaceae.

**78b Leaves usually compound, variously alternate, opposite, or seldom whorled; ovules two (seldom one or several) in each of the (2-) several locules; flowers with an evident nectary disk around the base of the ovary, or the disk modified into a gynophore ....
.... Citrus Family, RUTACEAE**

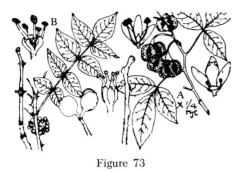

Figure 73

Figure 73   A, *Ptelea trifoliata* L., Common Hoptree; B, *Zanthoxylum americanum* Miller, Common Prickly-Ash, are both familiar species in the eastern United States.

Most of the 1,600 species of this family grow in warm countries. The citrus fruits are important members of the Rutaceae. *Phellodendron* (Cork-tree), *Dictamnus* (Gas-plant) and *Ruta* (Rue) are occasionally planted as ornamentals.

79a   (from 77b) Fruit a double samara, even the developing ovary flattened and showing the beginning of the two wings soon after pollination; leaves opposite, usually simple and palmately lobed, but sometimes compound; woody plants .... Maple Family, ACERACEAE

Figure 74

Figure 74   A, *Acer platanoides* L., Norway Maple; B, *Acer spicatum* Lam., Mountain Maple.

There are about 150 species of Aceraceae, nearly all of them in the genus *Acer*. Sugar Maple, *Acer saccharum* Marsh., is an important forest tree in the northeastern United States, valued for its wood and for the syrup and sugar made from its sap. Maples are a vivid part of the fall coloration in the northeastern United States, the leaves turning red or orange after a series of sunny days and cold nights. The woody habit, opposite, often palmately lobed leaves, and characteristic double samara make this family easily recognizable.

79b   Fruit of various sorts, but not a double samara; leaves and growth habit various 80

80a   (from 79b) Leaves tiny, scale-like, generally less than three mm long; halophytic or xerophytic, rather large shrubs or small trees with slender branches;

ovary unilocular .... Tamarisk Family,
TAMARICACEAE

Figure 75

Figure 75 *Tamarix gallica* L., Tamarisk.

Several species of *Tamarix* have been introduced into southwestern United States, where
they thrive along dry washes in the desert.
There are about a hundred species in the family, all native to the Old World.

80b **Leaves usually normally developed and
more than three mm long, but if tiny
and scale-like, then the plants herba-
ceous; ovary various 81**

81a (from 80b) Anthers opening by valves
uplifting from the base; stamens very
often in sets of three; ovary unilocular
.... 82

81b Anthers opening by slits or pores; stamens only seldom in sets of three; ovary
with one to several locules 83

82a (from 81a) Flowers perigynous, with a
well developed hypanthium resembling
a calyx tube; ovary with a single ovule
pendulous from near the summit; aro-
matic trees or shrubs ....
.... Laurel Family, LAURACEAE

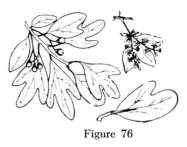

Figure 76

Figure 76 *Sassafras albidum* (Nutt.) Nees,
Sassafras, is sometimes called Mitten Tree because of the shape of some of its leaves, which
range from entire to mitten-shaped to three-
lobed. Spring tonic has traditionally been made
from the bark of the roots of Sassafras.

There are about 2,000 species of Lauraceae,
most of them tropical. *Laurus nobilis* L.,
Laurel, *Persea americana* Miller, Avocado,
Umbellularia californica Nutt., California
Laurel, *Lindera benzoin* (L.) Blume, Spice
Bush, *Cinnamomum zeylanicum* Blume, Cin-
namon, and *Cinnamomum camphora* (L.)
Nees & Eberm., Camphor, are some other fa-
miliar members of the Lauraceae.

82b **Flowers hypogynous, without a hypanthium or calyx tube; ovules usually several or many on a parietal placenta, or
seldom solitary and basal; herbs and
shrubs, usually not aromatic ....
Barberry Family, BERBERIDACEAE**

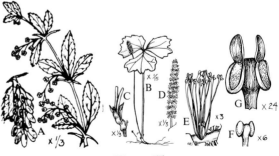

Figure 77

Figure 77 A, *Berberis vulgaris* L., Common Barberry, the alternate host of the fungus that causes stem rust (red rust, black rust) of wheat; B-G, *Achlys triphylla* (Smith) DC., Vanilla Leaf, a woodland species of the Pacific Northwest. B, habit; C, base of plant; D, inflorescence; E, single flower, with long stamens and short ovary; F, anther; G, anther, showing dehiscence by uplifted valves.

There are about 650 species of Berberidaceae, widespread especially in temperate parts of the Northern Hemisphere. Several species of *Berberis*, section *Mahonia* (sometimes considered to form a distinct genus *Mahonia*) are occasionally cultivated as ornamentals. They are evergreen shrubs with pinnately compound, spinulose-toothed, often shiny leaves and rather small, rich yellow flowers. *Berberis aquifolium* Pursh is Oregon Grape, the state flower of Oregon.

83a (from 81b) Stamens (including well developed staminodes) more than six 84

83b Stamens six or fewer 130

84a (from 83a) Ovary unilocular (sometimes with partial partitions extending inward from the sides but not joined in the center, or sometimes with partitions at the base, but these not reaching as high as the middle of the ovary) 85

84b Ovary with two or more locules, the partitions complete at least to above the middle of the ovary 104

85a (from 84a) Placentas two or more 86

85b Placenta only one (but often bearing more than one ovule) 93

86a (from 85a) Plants with milky or colored juice 87

86b Plants with colorless, non-milky juice 88

87a (from 86a) Sepals two or three; anthers opening by longitudinal slits; leaves without stipules, variously simple and entire to more often lobed or compound or dissected; mostly herbs .... Poppy Family, PAPAVERACEAE

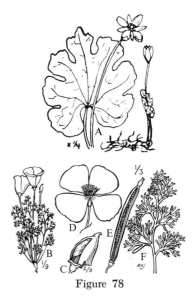

Figure 78

Figure 78 A, *Sanguinaria canadensis* L., Bloodroot; B-F, *Eschscholtzia californica* Cham., California Poppy, is the state flower of California. B, habit; C, flower bud opening and pushing off the sepals, which remain connate at the tip; D, flower; E, opening fruit; F, leaf.

There are about 200 species of Papaveraceae, which occur mainly in temperate and subtropical parts of the Northern Hemisphere. The genus *Papaver*, for which the family is named, has the partitions in the ovary so nearly complete that it is also keyed in another place. *Argemone* (Prickly Poppy, with large flowers and prickly-margined leaves) and *Chelidonium* (Celandine) are some other familiar genera of Papaveraceae. Very few families of herbaceous dicotyledons with well developed flowers and a single pistil have only two sepals. The principal other ones are the Fumariaceae, with irregular flowers, and the Portulacaceae, with free-central placentation. See also fig. 99.

87b Sepals five; anthers opening by terminal pores or short slits; leaves stipulate, simple and palmately veined to palmately compound; trees, shrubs, or herbs .... Lipstick-Tree Family, BIXACEAE

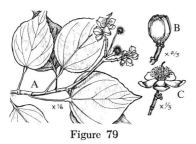

Figure 79

Figure 79 *Bixa orellana* L., Lipstick-Tree, is the source of an orange dye. A, habit; B, flower bud; C, flower.

There are only three genera and about twenty-five species of Bixaceae, all native to tropical America. The other two genera, *Amoreuxia* and *Cochlospermum*, have the ovary partitioned at the top and bottom, but not at the middle.

88a (from 86b) Styles two or more and distinct, or united only part-way to the summit, or the stigmas sessile but well separated on the ovary 89

88b Style solitary, with a simple or merely lobed stigma, or the stigmas sessile and

grouped together on the summit of the ovary 90

89a (from 88a) Herbs; stamens up to 10 in number; leaves variously simple and entire to lobed, compound, or dissected .... Saxifrage Family, SAXIFRAGACEAE

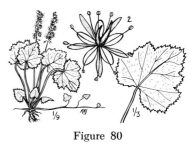

Figure 80

Figure 80 *Tiarella cordifolia* L., Foam Flower, is widespread in eastern United States, and similar species grow in the West.

There are about 700 species of Saxifragaceae, most of them in temperate and cold regions. The family shows an extraordinary diversity in floral structure, with two to four carpels that are sometimes separate nearly to the base, but usually united to varying degrees to form a compound ovary (often apically cleft) with axile or parietal placentation; the flowers range from virtually hypogynous to more often perigynous or sometimes virtually epigynous; petals are present or sometimes absent, and the stamens are mostly as many or twice as many as the calyx lobes. See also figs. 43, 109, 153.

89b Woody plants, mostly everygreen; stamens generally numerous; leaves simple and entire or merely toothed .... Flacourtia Family, FLACOURTIACEAE

Figure 81

Figure 81 *Hydnocarpus venenata* Gaertn., Chaulmoogra Tree.

There are about 1,200 species of Flacourtiaceae, most of them tropical, none in the United States. Chaulmoogra oil, obtained from species of *Hydnocarpus,* was formerly used in the treatment of leprosy.

90a **(from 88b) Sepals four; petals four; ovary often elevated on a gynophore; leaves often compound; fruit often with a replum (see lead no. 146) .... Caper Family, CAPPARACEAE**

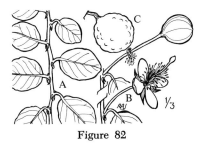

Figure 82

Figure 82 *Capparis spinosa* L., Caper-Bush. A, leafy twig; B, portion of inflorescence, showing flower at anthesis (below), and developing fruit with a long gynophore above the receptacle (above); C, fruit.

There are about 800 species of Capparaceae, most of them in tropical or subtropical regions, but only a limited number of them will

key here. Others have only six stamens and will come out at another place in the key. See figs. 140, 171.

90b **Sepals three or five; petals three or five; ovary sessile, not elevated on a gynophore; leaves simple; fruit without a replum 91**

91a **(from 90b) Filaments connate into a tube; fruit a berry; plants aromatic, glabrous, woody .... Canella Family, CANELLACEAE**

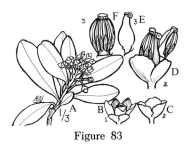

Figure 83

Figure 83 *Canella winterana* (L.) Gaertn., is the only species of this small (20 spp.), tropical family that grows in the United States. The species is basically West Indian, but its range extends to southern Florida. A, habit; B, flower, external view; C, calyx; D, flower, with petals removed and with the stamen-column laid open to show the pistil; E, pistil; F, stamen-column, with partially opened anthers. *Canella* bark has traditionally been used as a tonic and stimulant.

91b **Filaments separate; fruit a loculicidal capsule; plants not aromatic, often evidently hairy, variously woody or herbaceous 92**

92a **(from 91b) Anthers opening by pores at the apparent tip; stamens attached di-**

rectly to the receptacle; seeds very numerous and tiny, with minute, undifferentiated embryo .... **Shinleaf Family, PYROLACEAE**

Figure 84

Figure 84 *Pyrola secunda* L., One-sided Wintergreen.

There are about forty-five species of Pyrolaceae, mostly in the cooler parts of the Northern Hemisphere. Members of this family have a mycorhizal fungus that provides at least part of their food. The Pyrolaceae are evidently allied to the Ericaceae, and are often included in that family.

92b Anthers opening by longitudinal slits; stamens borne on a disk-like projection of the receptacle; seeds of moderate number, with well developed, usually curved or folded or coiled embryo .... **Rock-Rose Family, CISTACEAE**

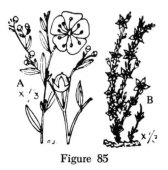

Figure 85

Figure 85 A, *Helianthemum canadense* Michx., Rock Rose; B, *Hudsonia ericoides* L., False Heather.

There are about 200 species of Cistaceae, half of them in the genus *Helianthemum*. The principal center of diversity for the family is in the Mediterranean region, but a score or so of species grow in eastern United States.

93a (from 85b) Styles two or more and distinct, or united only part way to the summit, or the style solitary but the stigmas clearly separate, or the stigmas sessile and distinct **94**

93b Style solitary, with a simple or merely lobed stigma, or the solitary stigma sessile **97**

94a (from 93a) Woody plants, usually with pinnately compound or trifoliolate leaves; fruit usually fleshy and indehiscent (a drupe) .... **Sumac Family, ANACARDIACEAE**

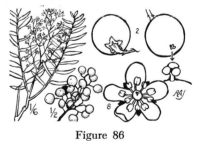

Figure 86

Figure 86 *Schinus molle* L., Pepper-tree, is native to tropical America, but often planted in the southern United States.

Some other species of Anacardiaceae have only five stamens and will key out at another place. See also fig. 127.

94b Herbs or seldom woody plants, always with simple leaves; fruit dry, variously dehiscent or indehiscent 95

95a (from 94b) Leaves opposite; ovules 1-many on a free-central or basal placenta .... Pink Family, CARYOPHYLLACEAE

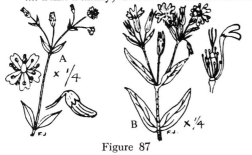

Figure 87

Figure 87   A, *Cerastium nutans* Raf., Nodding Chickweed; B, *Silene noctiflora* L., Night-flowering Catchfly.

There are about 2,000 species of Caryophyllaceae. The bulk of the family is marked by its combination of herbaceous habit, opposite leaves, four or five sepals (or a four- or five-lobed calyx), separate petals, and free-central or basal placentation. Carnation, Sweet William, and the various kinds of Pinks (all members of the genus *Dianthus*) belong to this family, as does Baby's Breath (*Gypsophila*), used in floral arrangements. See also fig. 149.

95b Leaves alternate (sometimes all basal) 96

96a (from 95b) Perianth of two (seldom several) distinct sepals and 4-6 (seldom more numerous) distinct, well differentiated petals; ovules generally several or many on a free-central placenta; fruit usually dehiscent ....
Purslane Family, PORTULACACEAE

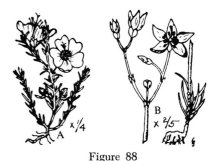

Figure 88

Figure 88   A, *Portulaca grandiflora* Hook., Rose Moss; B, *Talinum calycinum* Engelm., Large-flowered Talinum.

There are about 500 species of Portulacaceae, but many of them have only five stamens, instead of numerous stamens like the ones keyed here. *Lewisia rediviva* Pursh, the Bitterroot, is the state flower of Montana. It is unusual in its family in having several sepals, 12-18 petals, and numerous stamens. See figs. 142, 175.

96b Perianth of 2-6 tepals, these basally connate into a minute or evident floral tube, green and herbaceous to often colored and more or less petaloid, often in two similar or slightly dissimilar sets of three, or a single set of five; ovule solitary on a basal placenta; fruit dry and indehiscent .... Buckwheat Family, POLYGONACEAE

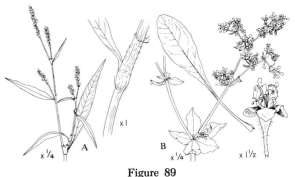

Figure 89

Figure 89 A, *Polygonum persicaria* L., Smart-weed; B, *Eriogonum tomentosum* Michx., Wild Buckwheat.

There are about 700 species of Polygonaceae, most of them native to the North Temperate Zone. Rhubarb (*Rheum*) and Buckwheat (*Fagopyrum*) are economically important members of the family. Many of the Polygonaceae, with the notable exception of the large genus *Eriogonum,* have a well developed, membranous stipular sheath, which immediately sets them apart from the vast majority of other dicotyledons. See also figs. 50, 148.

97a **(from 93b) Leaves with stipules, these sometimes developed into spines; nearly all woody plants** **98**

97b **Leaves without stipules; plants woody or herbaceous** **100**

98a **(from 97a) Fruit a pod, commonly opening along the two edges to release the several seeds; leaves simple, or more often pinnately or bipinnately compound; flowers hypogynous or slightly perigynous** .... **Mimosa Family, MIMOSACEAE**

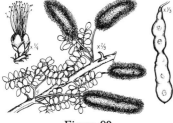

Figure 90

Figure 90 *Acacia greggii* A. Gray, Catclaw Acacia, a common desert species in the southwestern United States, showing a leafy twig with flower-spikes, plus a single flower and a fruit.

Most members of the Mimosaceae have the petals united below to form a tube, but some species of *Acacia* and other genera have separate petals. *A. greggii*, the species shown here, has the petals irregularly connate, or some of them distinct. See fig. 203.

98b **Fruit fleshy and indehiscent, a drupe with a single seed; leaves simple; flowers more or less strongly perigynous** **99**

99a **(from 98b) Style gynobasic; filaments in ours connate below** .... **Chrysobalanus Family, CHRYSOBALANACEAE**

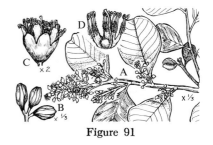

Figure 91

Figure 91 *Chrysobalanus icaco* L., Coco Plum. A, habit; B, fruits; C, flower; D, portion of flower showing irregularly connate stamens, ovary, and gynobasic style. Coco plum, native to southern Florida and much of the Caribbean region, is one of the few species of its family in the United States.

There are about 400 species of Chrysobalanaceae. The family is pantropical, but best developed in the New World. It is of no great economic importance.

99b Style terminal; filaments distinct ....
.... **Rose Family, ROSACEAE**

Figure 92

Figure 92 *Prunus nigra* Aiton, Canada Plum. Of the Rosaceae, only the large genus *Prunus* will key here. Nearly all other genera of the family either have more than one pistil or have an inferior ovary. The leaves and twigs of *Prunus* have the characteristic odor and taste of prussic acid. See figs. 27, 42, 182.

100a (from 97b) Plants woody; leaves simple; ovule mostly solitary **101**

100b Plants herbaceous; leaves compound or dissected, or sometimes simple and merely lobed **103**

101a (from 100a) Flowers hypogynous, without a hypanthium; ovule basal or nearly so **102**

101b Flowers distinctly perigynous, with a definite hypanthium; ovule pendulous from near the summit of the locule .... **.. Daphne Family, THYMELAEACEAE**

Figure 93

Figure 93 *Dirca palustris* L., Leatherwood. Indians used the tough, yellow-green stems of Leatherwood as cords for tying.

There are about 500 species of Thymelaeaceae, most of them in the Old World. Species of *Daphne* are grown as garden ornamentals.

102a (from 101a) Plants aromatic; flowers unisexual; filaments united into a solid column .... **Nutmeg Family, MYRISTICACEAE**

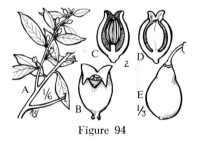

Figure 94

Figure 94 *Myristica fragrans* Houttuyn, Nutmeg, is native to the Molucca Islands, in the southwest Pacific. A, habit; B, external view of staminate flower; C, cut away view of staminate flower, showing column of stamens; D, cut away view of pistillate flower, showing ovary; E, fruit.

There are nearly 300 species of Myristicaceae, all tropical. Only the Nutmeg is of much economic importance.

102b Plants not aromatic; flowers usually perfect; filaments distinct, or connate into a tube around the ovary at the base .... Four-O'Clock Family, NYCTAGINACEAE

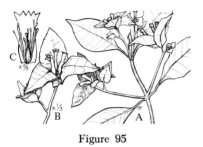

Figure 95

Figure 95 *Bougainvillea glabra* Choisy is a South American woody vine that is often cultivated in door-yards in southern United States. The showy, usually pink-purple bracts subtend small clusters of less showy flowers. A, habit; B, detail of flower cluster; C, flower, laid open to show stamens and pistil.

Some flowers of *Bougainvillea,* and all those of our other members of the Nyctaginaceae, have six or fewer stamens and will key to another place. See figs. 143, 164.

103a (from 100b) Leaves alternate, twice or thrice ternately compound .... Buttercup Family, RANUNCULACEAE

Figure 96

Figure 96 *Actaea alba* (L.) Mill., Doll's Eyes, has shiny white berries. The related species *Actaea rubra* (Aiton) Willd. has bright red berries. Species of *Actaea* have a characteristic pungent odor.

Only *Actaea* will key here. Nearly all other genera of the large family Ranunculaceae have more than one pistil and will key elsewhere. See fig. 39.

103b Leaves opposite, deeply lobed, but not compound .... Barberry Family, BERBERIDACEAE

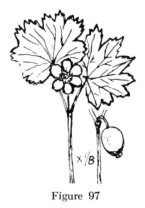

Figure 97

Figure 97 *Podophyllum peltatum* L., May-Apple, is a common early-blooming herb in the forests of eastern United States.

Other members of the Berberidaceae have more specialized anthers and will key elsewhere. See fig. 77.

104a (from 84b) Stamens numerous, more than 12 105

104b Stamens 7-12 114

105a (from 104a) Plants succulent; petals generally numerous and narrow; herbs or shrubs .... Fig-Marigold Family, AIZOACEAE

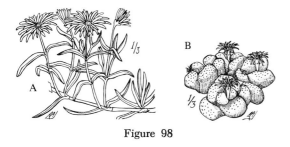

Figure 98

Figure 99

Figure 98 A, *Lampranthus* (*Mesembryanthemum*) *multiradiatus* (Jacq.) N. E. Brown, Fig-Marigold; B, *Lithops localis* (N. E. Brown) Schwantes, Stone-Plant.

There are about 2,500 species of Aizoaceae, the vast majority of them native to South Africa or Australia. Some of them are grown in greenhouses or as curiosities in homes. Leaf-succulents with numerous, narrow petals mostly belong to the Aizoaceae. The ovary ranges from superior to inferior and is usually 2-5-locular. Most of the species of Aizoaceae have traditionally been included in the very large and unwieldy genus *Mesembryanthemum*, but more recent botanists usually see 50 to 100 (or even more) smaller, closely related genera in its place.

105b Plants not notably succulent; petals generally four or five, or up to eight (more numerous in some cultivated double forms, but then broad) **106**

106a (from 105b) Sepals two; petals usually four (more numerous in doubles); herbs .... **Poppy Family, PAPAVERACEAE**

Figure 99 *Papaver rhoeas* L., Corn Poppy. Unlike other members of its family, *Papaver* has the parietal placentas so deeply intruded as partial partitions that the placentation may at first inspection appear to be axile. Opium is obtained from the milky juice of *Papaver somniferum* L., and some other poppies contain similar narcotics. For other genera in the Papaveraceae, see fig. 78.

106b Sepals (3) 4 or 5 (-8); petals usually 5 (-8) **107**

107a (from 106b) Flowers strongly perigynous, with a more or less cup-shaped hypanthium that has the sepals (like calyx-lobes), petals, and sometimes also the stamens attached around its summit; leaves all or mainly opposite or whorled (the upper ones sometimes alternate) .... **Loosestrife Family, LYTHRACEAE**

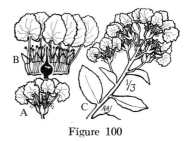

Figure 100

Figure 100 *Lagerstroemia indica* L., Crape Myrtle, is a commonly planted small tree in southeastern United States. A, external view of flower; B, flower, laid open, with numerous stamens, and with six petals attached to the hypanthium, each petal with a long, slender claw and a broad, ruffled blade; C, flowering twig.

Most members of the Lythraceae are herbaceous and have relatively few stamens, and thus will key elsewhere. See figs. 49, 108, 151.

107b Flowers hypogynous or nearly so; leaves alternate **108**

108a (from 107b) Sepals or calyx-lobes imbricate, the margins overlapping in bud; plants glabrous or hairy, but without stellate hairs or peltate scales; leaves usually pinnately veined **109**

108b Sepals or calyx-lobes valvate, the margins merely edge to edge in bud, not overlapping; plants very often with stellate hairs or small, peltate scales; leaves very often palmately veined, sometimes palmately lobed or even compound .. **111**

109a (from 108a) Leaves with stipules; style very often gynobasic, the ovary so deeply lobed that its lobes appear to be united only by the common style ....
.... **Ochna Family, OCHNACEAE**

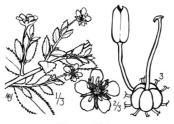

Figure 101

Figure 101 *Ochna multiflora* DC., an African species sometimes cultivated in warm countries as a street tree.

There are about 400 species of Ochnaceae, many of them in Brazil.

109b Leaves without stipules; styles terminal, the ovary not deeply lobed **110**

110a (from 109b) Anthers more or less erect, not inverted, opening fully by longitudinal slits; trees or shrubs ....
.... **Tea Family, THEACEAE**

Figure 102

Figure 102 *Stewartia ovata* (Cav.) Weatherby, Mountain Stewartia.

There are about 600 species of Theaceae, most of them in warm regions. Tea is made from the leaves of *Thea sinensis* L. The Franklin Tree (*Franklinia alatamaha* Marsh.), originally native in southeastern United States, is frequently cultivated but has not been seen in the wild for many years. *Camellia* is another familiar genus of the Theaceae. It is the state flower of Alabama.

110b Anthers inverted, opening by seemingly terminal pores (in *Actinidia* the conspicuous, broad pore eventually prolonged below into a narrow slit); trees, shrubs, or often woody vines ....
.... **Chinese Gooseberry Family, ACTINIDIACEAE**

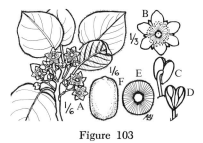

Figure 103

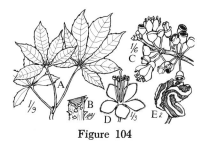

Figure 104

Figure 103 *Actinidia chinensis* Planch., Chinese Gooseberry. A, flowering branch; B, flower, from above; C, D, back and front view of inverted, poricidal anther; E, cross section of fruit, with numerous carpels; F, external view of fruit. The edible fruit is sometimes available in specialty markets.

There are about 300 species of Actinidiaceae, most of them in tropical and subtropical countries. The largest genus is *Saurauia,* which grows in tropical Asia as well as in tropical America.

111a (from 108b) Anthers unilocular (with a single pollen sac, opening by a single slit); flowers very often with an epicalyx (narrow, sepal-like parts attached alternately with the larger and broader sepals or calyx-lobes) 112

111b Anthers bilocular (with two pollen sacs, each opening by a slit); flowers usually without an epicalyx; mostly trees and shrubs 113

112a (from 111a) Trees; pollen generally smooth or merely finely ridged on the surface, with three germ pores; filaments all connate into a tube around the ovary, or often connate into several separate phalanges .... **Kapok Tree Family, BOMBACACEAE**

Figure 104 *Ceiba pentandra* (L.) Gaertn., Kapok Tree, is a widespread tropical species. A, twig with leaves; B, spiny stem; C, leafless flowering twig; D, flower; E, compound anther, representing the pollen-sacs of several stamens atop a compound common filament.

There are about 200 species of Bombacaceae, most of them tropical; they are especially abundant in tropical America. *Ochroma pyrimadale* (Cav.) Urban, of tropical America, is the source of balsa wood. The famous Baobab Tree of tropical Africa, *Adansonia digitata* L., is another member of the Bombacaceae. The Bombacaceae are so closely related to the Malvaceae that botanists find it difficult to draw a clear line between the two groups.

112b Herbs or soft-wooded shrubs; pollen generally spinulose and with several or many germ pores; filaments all connate into a tube around the ovary ....
.... **Mallow Family, MALVACEAE**

Figure 105

Figure 105 A, *Hibiscus trionum* L., Flower-of-an-Hour; B, *Malva rotundifolia* L., Cheeses. Both of these species are European plants that are widely introduced in the United States. Several species of *Hibiscus* have spectacular large flowers, and one species, *H. esculentus* L., is cultivated for its edible fruits (okra). *Hibiscus* is the state flower of Hawaii.

There are about 1,500 species of Malvaceae, the majority of them tropical or subtropical. Herbs with the stamens united by their filaments to form a tube around a compound ovary with separate styles (or style branches) mostly belong to the Malvaceae. Cotton is made from the seed-hairs of species of *Gossypium*. The Garden Hollyhock, *Althaea rosea* (L.) Cav., is another well known member of the family.

**113a (from 111b) Filaments generally all connate into a tube around the ovary ....
...... Cacao Family, STERCULIACEAE**

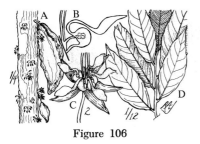

Figure 106

Figure 106 *Theobroma cacao* L., Cacao. A, portion of branch of tree, with clusters of flowers and a single fruit; B, detail of hooded petal enclosing the anther of a stamen; C, flower; D, leafy twig. Cacao is native to tropical America, but it is also cultivated in Africa. Cocoa and chocolate are made from the seeds. The flower has five anther-bearing stamens and five long, slender staminodes.

There are about a thousand species of Sterculiaceae, most of them native to tropical and subtropical regions. The largest genera are *Sterculia* and *Dombeya*. Species of the African genus *Cola* provide caffeine and colanine (a heart stimulant) for use in beverages. The Sterculiaceae, Bombacaceae and Malvaceae have valvate sepals (or calyx-lobes) and very often have stellate pubescence. See also fig. 123.

**113b Filaments distinct, or very shortly connate into bundles at the base ....
.... Basswood Family, TILIACEAE**

Figure 107

Figure 107. *Tilia americana* L., American Basswood, is a common member of the climax forest in much of the northeastern United States. The common name Linden, originally used for some European species of *Tilia*, is sometimes also applied to our native species. The slender, somewhat leaf-like bract on the peduncle is a characteristic feature of *Tilia*.

There are about 400 species of Tiliaceae, most of them in warm regions. *Tilia* is the most familiar genus in the North Temperate Zone.

114a (from 104b) Herbs, or seldom low shrubs or half-shrubs 115

114b Shrubs or trees, or sometimes woody vines 123

115a (from 114a) Flowers distinctly perigynous, with a well developed hypanthium 116

115b Flowers hypogynous, without a hypanthium 118

116a (from 115a) Leaves mostly opposite or whorled, only rarely alternate; style solitary, elongate, with a mostly capitate stigma ....
...... Loosestrife Family, LYTHRACEAE

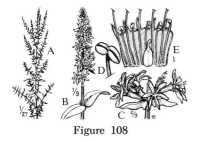

Figure 108

Figure 108 *Lythrum salicaria* L., Purple Loosestrife, is a showy European species that has become well established in moist or wet open places in eastern United States. It is unusual in being tristylic; i.e., there are three different modes of relative length of the style and stamens in flowers of different plants. This complex arrangement favors cross-pollination, even though the flowers are perfect. A, habit; B, portion of inflorescence; C, cluster of flowers at a node; D, anther and upper part of filament; E, flower laid open, with the petals removed, showing pistil, short and long stamens, and hypanthium wth calyx-lobes.

There are nearly 500 species of Lythraceae, most of them tropical. *Lawsonia inermis* L., widely cultivated in the tropics, is the source of henna. See also figs. 49, 100, 151.

116b Leaves alternate, or all basal, only rarely opposite; styles two or more and distinct, or sometimes connate toward the base 117

117a (from 116b) Styles and locules two; stamens 10; sepals five; plants not aromatic .... Saxifrage Family, SAXIFRAGACEAE

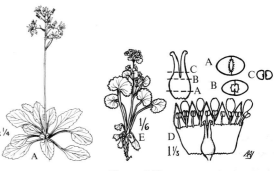

Figure 109

Figure 109 A, *Saxifraga virginiensis* Michx., Virginia Saxifrage; B-E, *Telesonix jamesii* (Torr.) Raf., a Rocky Mountain species. B, C, external and sectional views of pistil; D, flower, laid open; E, habit.

There are about 700 species of Saxifragaceae, but many of them have a unilocular ovary with parietal placentation, and will therefore key to another place. See figs. 43, 80, 153.

117b Styles and locules six; stamens 12; sepals three; plants aromatic .... Birthwort Family, ARISTOLOCHIACEAE

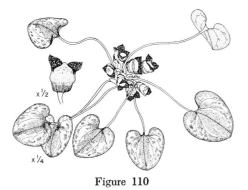

Figure 110

Figure 110 *Hexastylis virginica* (L.) Small, Virginia Heart-Leaf, is a common herb of the forest floor in the southern Appalachian region. Other genera of the Aristolochiaceae have an inferior ovary and will key elsewhere. See fig. 176.

118a (from 115b) Styles several and separate **119**

118b Style 1, sometimes cleft above 120

119a (from 118a) Leaves simple, entire; carpels more than five; ovules solitary in each locule; embryo marginal in the seed, curved around the perisperm .... Pokeweed Family, PHYTOLACCACEAE

Figure 111

Figure 111 *Phytolacca americana* L., Common Pokeweed, is common over most of the eastern United States. The small flowers are whitish, and the pea-sized berries are nearly black, with purple juice. The roots are highly poisonous, but the young stems and leaves can be eaten as cooked greens after their poison has been extracted in boiling water.

There are about 135 species of Phytolaccaceae, many of them in the warmer parts of the New World.

119b Leaves compound; carpels mostly five; ovules 2-several in each locule; embryo straight, embedded in the endosperm .... Wood Sorrel Family, OXALIDACEAE

Figure 112

Figure 112 *Oxalis stricta* L., Yellow Wood Sorrel, is a dooryard weed in most of the United States. Many people believe that the original Shamrock of Ireland was a species of *Oxalis*. Others are equally sure that it was *Trifolium repens* L., the White Clover, of the Pea Family.

There are nearly a thousand species of Oxalidaceae, most of them in tropical or subtropical latitudes, but often at high altitudes.

120a (from 118b) Ovules and seeds very numerous in each of the five locules of the ovary .... Shinleaf Family, PYROLACEAE

See fig. 84, *Pyrola secunda* L. The ovary of *Pyrola* appears on first inspection to be five-locular, but the partial partitions are not actually joined in the center; thus the ovary is technically unilocular with deeply intruded placentas.

120b Ovules and seeds several (but not very numerous) in each of the five locules of the ovary **121**

121a (from 120b) Style gynobasic, uniting the otherwise essentially distinct carpels; leaves alternate, pinnately parted or compound or dissected .... **Meadow-Foam Family, LIMNANTHACEAE**

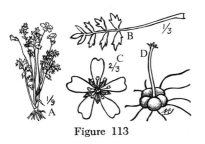

Figure 113

Figure 113 *Limnanthes douglasii* R. Br., is native to California as are most other species of *Limnanthes*. A, habit; B, leaf; C, flower, from above; D, pistil, showing deeply lobed ovary and gynobasic style. The family Limnanthaceae has only the genera *Limnanthes* and *Floerkea*, about a dozen species in all. Species of *Limnanthes* are attracting some interest as a potential new agricultural crop, because of the unusual oil in their seeds. For *Floerkea*, see fig. 154.

121b Style terminal on the ovary **122**

122a (from 121b) Leaves alternate, or all basal, variously palmately lobed or compound, or odd-pinnately lobed or compound; lower part of the style forming a prominently thickened, tapering column .... **.... Geranium Family, GERANIACEAE**

Figure 114

Figure 114 *Geranium maculatum* L., Wild Geranium, is native to the eastern United States, but several similar species are widespread in the West.

There are about 700 species of Geraniaceae, most of them in temperate and warm-temperate regions. The largest genera are *Geranium* and *Pelargonium*, the latter known in cultivation also as Geranium. See also figs. 155, 161.

122b Leaves opposite, even-pinnately compound, without a terminal leaflet; style slender, not forming a thickened column .... **Creosote Bush Family, ZYGOPHYLLACEAE**

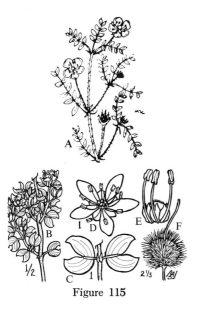

Figure 115

123b Leaves mostly or all alternate, simple or compound, but if compound then odd-pinnate, with a terminal leaflet 124

124a (from 123b) Leaves simple; plants unarmed, not thorny 125

124b Leaves compound, except for a few Simaroubaceae that have thorny branches 127

125a (from 124a) Anthers opening by small, slit-like pores at the apparent tip; fruit a capsule with numerous seeds ....
.... White Alder Family, CLETHRACEAE

Figure 115 A, *Kallstroemia intermedia* Rydb., Greater Caltrop; B-F, *Larrea tridentata* (DC.) Coville, Creosote Bush, is the dominant plant in much of the desert region of southwestern United States and northern Mexico. Its small, hard, shiny leaves can withstand severe desiccation. B, flowering twig; C, pair of leaves, each with two leaflets; D, flower; E, portion of flower, showing stamens with basal appendages; F, ovary and style. Puncture Weed, *Tribulus terrestris* L., is another member of this family; its small, hard, spiny fruits are notorious for their effect on bicycle tires and bare feet.

There are about 250 species of Zygophyllaceae, most of them native to warm, dry regions. Some are herbs, and will key here; others are woody plants and will key out under lead 123a.

123a (from 114b) Leaves mostly or all opposite, even-pinnately compound, without a terminal leaflet .... Creosote Bush Family, ZYGOPHYLLACEAE
See fig. 115.

Figure 116

Figure 116 *Clethra alnifolia* L., White Alder, is a shrub of eastern United States.
The family Clethraceae, related to the Heath Family, has only the genus *Clethra*, with some thirty or forty species.

125b Anthers opening by longitudinal slits that run their whole length; fruit indehiscent, 1-seeded 126

126a (from 125b) Petals with a tongue-like appendage on the inner side toward the base; fruit a drupe, not winged ....
.. Coca Family, ERYTHROXYLACEAE

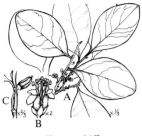

Figure 117

Figure 117 *Erythroxylon coca* Lam., Coca Plant, the source of cocaine, is a shrub of the South American Andes. A, habit; B, flower; C, developing fruit.

There are about 200 species in the family, nearly all of them in the single large genus *Erythroxylon,* which centers in South America.

126b Petals not appendaged; fruit dry, subtended by a persistent calyx, usually two, three, or all five of the calyx lobes accrescent and wing-like .... Dipterocarp Family, DIPTEROCARPACEAE

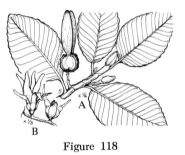

Figure 118

Figure 118 *Dipterocarpus alatus* Roxb. A, habit, with mature fruit and accrescent sepals; B, twig with flowers.

There are about 400 species of Dipterocarpaceae, nearly all of them in the Old World tropics, where they occur especially in rain forests of the Malaysian region. Many of them are valuable timber trees.

127a (from 124b) Nectary disk extrastaminal (outside the stamens); ovules apotropous, typically one in each locule and ascending-erect, with the micropyle turned outwards .... Soapberry Family, SAPINDACEAE

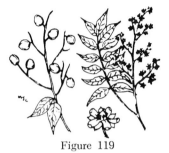

Figure 119

Figure 119 *Sapindus drummondii* Hook. & Arn., Western Soapberry, is a small tree native to southwestern United States.

There are about 1,500 species of Sapindaceae, most of them in warm regions. *Koelreuteria paniculata* Laxm., the Golden Rain Tree, is well known in cultivation in temperate climates.

127b Nectary disk intrastaminal, or developed into a gynophore or androgynophore; ovules epitropous (i.e., the micropyle turned toward the axis, and the raphe external, when the ovule is erect or ascending) 128

128a (from 127b) Plants strongly resinous, with intercellular resin canals in the bark and usually also in the wood rays .... Frankincense Family, BURSERACEAE

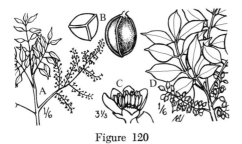

Figure 120

Figure 120 *Bursera simarouba* L., the Naked Indian, is conspicuous in Mexico and the West Indies and southernmost Florida because of its smooth, lustrous, coppery-red bark. A, habit, with flowers and developing young leaves; B, end and side views of fruit; C, flower, with one petal cut away; D, habit, with fruit and mature leaves.

There are about 600 species of Burseraceae, all of them in warm regions. Frankincense is obtained from *Boswellia carteri* Birdw., and myrrh from *Commiphora abyssinica* (Berg) Engl., both of this family.

128b Plants not strongly resinous **129**

129a (from 128b) Stamens distinct; seeds without endosperm .... **Quassia Family, SIMAROUBACEAE**

Figure 121

Figure 121 *Ailanthus altissima* (Miller) Swingle, Tree-of-Heaven, a native of central Asia, is now widely planted and spontaneous in cities of the United States. It is the tree of "A Tree Grows in Brooklyn."

There are about 150 species of Simaroubaceae, most of them tropical. The bitter bark has medicinal qualities.

129b Stamens usually connate by their filaments; seeds usually with endosperm .... **Mahogany Family, MELIACEAE**

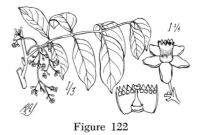

Figure 122

Figure 122 *Swietenia mahogani* Jacq., Mahogany, is a native of the West Indies, prized for use in fine furniture.

There are about 1,400 species of Meliaceae, most of them in warm regions. *Melia azedarach* L., China-Berry, is often cultivated as a street tree in southern United States. See also fig. 131.

130a (from 83b) Trees, shrubs (sometimes dwarf), or woody vines **131**

130b Herbs, or seldom half-shrubs **146**

131a (from 130a) Stamens opposite the petals (when these are present), alternate with the sepals **132**

131b Stamens alternate with the petals (when these are present), opposite the sepals, sometimes fewer than the number of petals or sepals 134

132a (from 131a) Petals none; sepals large and showy; ovary with many ovules in each locule .... Cacao Family, **STERCULIACEAE**

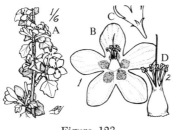

Figure 123

Figure 123 *Fremontia californica* Torr., Flannel Bush, is a showy flowering shrub of southern California. A, habit; B, flower; C, opening bud, showing sepal-like bracteoles below the corolla-like calyx; D, stamen-tube, surrounding the exserted style. Most members of the Sterculiaceae have two sets of stamens and will key to another place. See also fig. 106.

132b Petals present; sepals small and relatively inconspicuous; ovary with one or two ovules in each locule 133

133a (from 132b) Flowers perigynous; ovules solitary in each locule; fruit various, most commonly drupaceous, but not a berry; mostly shrubs or trees, often thorny, only rarely tendril-bearing vines; leaves simple, pinnately or palmately veined (species of *Forsellesia*, in the Crossosomataceae, might be sought here, except for the follicular fruit; see fig. 29) .... Buckthorn Family, **RHAMNACEAE**

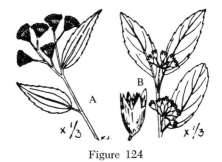

Figure 124

Figure 124 A, *Ceanothus americanus* L., New Jersey Tea; B, *Rhamnus caroliniana* Walter, Carolina Buckthorn.

There are about 900 species of Rhamnaceae, found throughout most of the world. The family is marked by its perigynous flowers with the stamens as many as and opposite the separate petals. Extracts from the bark and fruit of species of *Rhamnus* are used as laxatives. Species of *Ceanothus* are a common component of chaparral communities in western United States.

133b Flowers hypogynous; ovules two in each locule; fruit a berry; mostly tendril-bearing vines; leaves simple (very often palmately lobed or veined) or compound .... Grape Family, **VITACEAE**

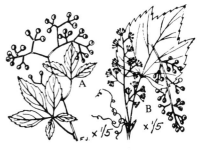

Figure 125

Figure 125   A, *Parthenocissus quinquefolia* (L.) Planch., Virginia Creeper, a decorative, high-climbing woody vine; B, *Vitis vulpina* L., Frost Grape.

There are about 700 species of Vitaceae, only a few of them in temperate climates. The family is marked by its mostly viny habit and hypogynous flowers with the stamens as many as and opposite the separate petals. *Vitis vinifera* L. is the grape traditionally used for making wine.

134a (from 131b) Ovary with a single locule and one or two ovules, or sometimes with one or more small empty locules in addition to the fertile one 135

134b Ovary with 2-5 (seldom more) locules, and 1-many ovules in each locule 139

135a (from 134a) Flowers strongly perigynous, with an elongate hypanthium resembling a calyx-tube; leaves and young twigs often covered with small scales or stellate hairs .... Russian Olive Family, **ELAEAGNACEAE**

Figure 126

Figure 126   *Elaeagnus commutata* Bernh., Silverberry, is a shrub with leaves, fruit, and outside of the flowers covered with small, peltate scales that give the surface a silvery appearance.

The Russian Olive, *Elaeagnus angustifolia* L., and the Buffalo Berry, *Shepherdia canadensis* (L.) Nutt., are some other familiar members of this rather small (50 spp.) family.

135b Flowers hypogynous or nearly so, without a prominent hypanthium; leaves and twigs only seldom provided with scales or stellate hairs 136

136a (from 135b) Leaves usually compound, only seldom simple; plants strongly resinous, with well developed resin-ducts in the bark and in the larger veins of the leaves .... Sumac Family, **ANACARDIACEAE**

Figure 127

Figure 127 A-D, *Toxicodendron radicans* (L.) Kuntze, Poison Ivy, is a climbing vine (or sometimes a slender shrub) with trifoliate leaves, the central leaflet on an evident stalk. A, habit, in flower; B, functionally pistillate flower; C, staminate flower; D, fruits and leaves; E, *Rhus glabra* L., Smooth Sumac, is a coarse shrub with pinnately compound leaves.

There are about 600 species of Anacardiaceae, most of them tropical. Many are poisonous to the touch, but others, such as the Mango (*Mangifera indica* L.) and Pistachio Nut (*Pistacia vera* L.) are edible. See also fig. 86.

136b Leaves simple; plants not strongly resinous, without resin-ducts **137**

137a (from 136b) Ovule or ovules pendulous from the top of the ovary **138**

137b Ovule basal, solitary .... **Goosefoot Family, CHENOPODIACEAE**

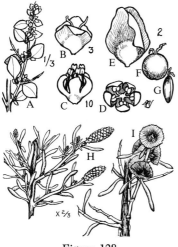

Figure 128

Figure 128 A-G, *Atriplex confertifolia* (Torr. & Frem.) S. Wats., Shadscale, is a common shrub of alkaline soils in the arid West. A, habit; B, bracts enclosing the pistillate flower, with two exserted styles; C, D, staminate flower, side and top views; E, fruiting pistillate bracts; F, G, two views of achene. H, I, *Sarcobatus vermiculatus* (Hook.) Torr., Greasewood, grows in even more alkaline sites, often in heavy soils with poor drainage. H, habit, with fleshy, ament-like staminate inflorescences and axillary pistillate flowers; I, twig with fruits and wing-like calyx.

Most members of the Chenopodiaceae are herbaceous and will key in another place. See fig. 147.

138a (from 137a) Leaves without stipules; petals usually present; ovules two .... **Icacina Family, ICACINACEAE**

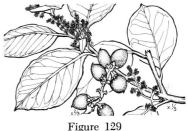

Figure 129

Figure 129 *Citronella gongonha* (Mart.) Howard, of Brazil, Uruguay, and Paraguay, is sometimes used as a substitute for tea.

There are about 400 species of Icacinaceae, nearly all of them tropical.

138b Leaves with stipules; petals none; ovule solitary ...... **Elm Family, ULMACEAE**

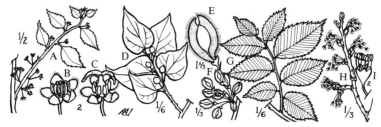

Figure 130

Figure 130   A-D, *Celtis occidentalis* L., Hackberry, provides good food for birds. A, twig with staminate flowers toward the base and pistillate flowers in the axils of the young leaves; B, staminate flower; C, functionally pistillate flower; D, twig with mature leaves and fruits. E-I, *Ulmus americana* L., American Elm, is a graceful tree that is now seriously threatened by disease. E, single fruit (samara); F, cluster of fruits; G, leafy twig; H, flowering twig in early spring, before the leaves develop; I, flower, with exserted stamens. Note that the leaves of *Celtis* have three main veins from the base, whereas the leaves of *Ulmus* are strictly pinnately veined.

There are about 150 species of Ulmaceae, most of them in the Northern Hemisphere.

139a (from 134b) Leaves compound 140

139b Leaves simple, entire or merely toothed **141**

140a (from 139a) Leaves alternate; stamens distinct, or often connate by their filaments; ovules mostly one or two in each locule ....
.... **Mahogany Family, MELIACEAE**

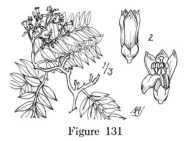

Figure 131

Figure 131   *Cedrela odorata* L., of the West Indies, is used in making furniture.

Most of the 1400 species of Meliaceae differ from *Cedrela* in having connate filaments, and many of them have more than six stamens and would thus key to another place. See fig. 122.

140b Leaves opposite; stamens distinct; ovules mostly 6-12 in two rows in each locule on the axile placentas (*Fraxinus*, in the Oleaceae, might be sought here, except that it has only two ovules per locule; See fig. 56) ....
.... **Bladdernut Family, STAPHYLEACEAE**

Figure 132

Figure 132   *Staphylea trifolia* L., American Bladdernut.

There are about fifty species of Staphyleaceae. *Staphylea,* the only genus in the United States, has a conspicuously inflated, few-seeded capsule that opens at the tip.

141a (from 139b) Leaves small, rarely as much as 1.5 cm long; low shrubs .... **Crowberry Family, EMPETRACEAE**

Figure 133

Figure 133   *Empetrum nigrum* L., Crowberry. A, leafy twig, with fruits; B, flower, with exserted stamens; C, flower, with perianth and bracteoles removed; D, fruit.

There are only five species of Empetraceae, irregularly distributed in temperate and cold regions. In addition to *Empetrum,* we have *Corema* near the coast in northeastern United States, and *Ceratiola* on the southeastern coastal plain.

141b Leaves larger, most or all of them more than 1.5 cm long; shrubs or trees **142**

142a (from 141b) Ovules more or less numerous; seeds with copious endosperm and tiny embryo .... **Pittosporum Family, PITTOSPORACEAE**

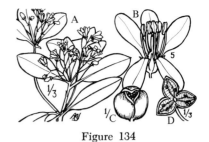

Figure 134

Figure 134   *Pittosporum tobira* Aiton, an oriental species cultivated in mild climates for its ornamental foliage. A, habit; B, flower; C, opening fruit; D, fully opened fruit.

There are about 200 species of Pittosporaceae, all native to the Old World, especially Australia. *Pittosporum* is by far the largest genus.

142b Ovules 1-2 (3) per locule, or up to 10 per locule in Hippocrateaceae; seeds various **143**

143a (from 142b) Ovules erect and basal-axile, or (most Hippocrateaceae) several and superposed in two rows on the axile placenta in each locule **144**

143b Ovules pendulous from the top of the ovary **145**

144a (from 143a) Stamens four or five; disk intrastaminal, or the stamens seated on the disk .... **Bittersweet Family, CELASTRACEAE**

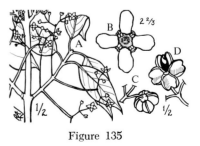

Figure 135

Figure 135 *Euonymus atropurpureus* Jacq., Burning Bush, an eastern American shrub cultivated for its foliage, which turns bright red in the fall. A, habit; B, tetramerous flower; C, ripening fruit; D, opened fruit, one seed still remaining.

There are about 800 species of Celastraceae, more of them native to tropical than to temperate regions. *Celastrus scandens* L., American Bittersweet, is another familiar member of this family. Its seeds are covered with a soft, thin, bright red aril.

144b Stamens mostly three; disk extrastaminal (outside of the stamens) .... Hippocratea Family, HIPPOCRATEACEAE

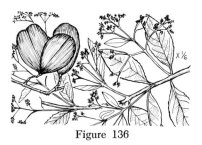

Figure 136

Figure 136 *Hippocratea volubilis* L., a native of tropical America, with wing-like fruits at left.

There are about 300 species of Hippocrateaceae, widespread in tropical regions. None is of any great economic importance. *Hippocratea* and *Salacia* are the largest genera.

145a (from 143b) Flowers perfect, borne in slender racemes; nectary disk present around the ovary .... Cyrilla Family, CYRILLACEAE

Figure 137

Figure 137 *Cyrilla racemiflora* L., Leatherwood, a shrub native to southeastern United States. (The common name Leatherwood is also applied to *Dirca palustris*, in the Thymelaeaceae.)

There are only fourteen species of Cyrillaceae, all in the New World. Only *Cyrilla* reaches the United States. *Cyrilla* is similar in aspect to *Clethra*, in the related family Clethraceae.

145b Flowers unisexual, borne in small cymes, or solitary; disk wanting .... Holly Family, AQUIFOLIACEAE

Figure 138

Figure 138   *Ilex opaca* Aiton, American Holly. There are three or four hundred species of *Ilex,* and only a few species in the other two genera of Aquifoliaceae. Several species of *Ilex* are cultivated for their ornamental foliage and bright red berries; other species have black fruits.

146a (from 130b) Ovary and fruit with a replum (a persistent, framelike parietal placenta, from which the valves of the fruit fall away at maturity); stamens generally six, but the sepals usually only four; petals usually four **147**

146b Ovary and fruit without a replum; stamens mostly five or fewer, but sometimes six, the sepals then only seldom four **148**

147a (from 146a) Replum forming a thin cross-partition in the ovary and fruit; stamens mostly tetradynamous, the two outer shorter than the four inner; ovary usually sessile on the receptacle, only seldom elevated on a gynophore .... **...... Mustard Family, BRASSICACEAE**

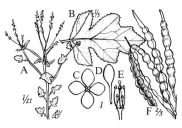

Figure 139

Figure 139   *Brassica nigra* L., Black Mustard, is a common field weed; table mustard is made from its seeds. A, habit of upper part of plant; B, leaf; C, flower, from above, showing cross-like form; D, a single petal; E, flower, stripped of its sepals and petals, showing pistil and short and long stamens; F, fruits, the one on the left showing the valves falling away from the persistent replum.

There are about 3,000 species of Brassicaceae (also called Cruciferae), mostly in the Northern Hemisphere. The family is marked by both its characteristic flowers (four sepals, four separate petals, two + four stamens) and characteristic fruit. Broccoli, Brussels Sprouts, Kale, Cabbage and Cauliflower are all forms of *Brassica oleracea* L.

147b Replum not forming a partition, the ovary unilocular; stamens all of about the same length; ovary usually elevated above the receptacle on a gynophore or androgynophore .... **........... Caper Family, CAPPARACEAE**

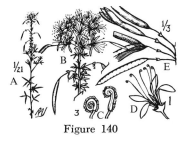

Figure 140

Figure 140   *Cleome serrulata* Pursh, Rocky Mountain Bee-plant. A, habit; B, inflorescence; C, anthers; D, flower, with four petals and six equal stamens on long filaments; E, fruits, with deciduous valves and persistent replum.

There are about 800 species of Capparaceae, most of them in warm, dry climates. The Capparaceae are closely related to the Brassicaceae and, like them, produce mustard oils. See also figs. 82, 171.

148a (from 146b) Ovary unilocular, and with basal or free-central placentation 149

148b Ovary either with more than one locule, or with two or more parietal placentas 157

149a (from 148a) Sepals two; petals five (seldom more); stamens mostly five and opposite the petals 150

149b Sepals three or more, or the perianth of two similar sets of sepaloid tepals; stamens variously disposed; flowers without definite petals except in some Caryophyllaceae, but commonly with a corolloid calyx in Nyctaginaceae 151

150a (from 149a) Plants twining or scrambling; ovule solitary; fruit indehiscent; leaves alternate .... Basella Family, BASELLACEAE

Figure 141

Figure 141 *Boussingaultia gracilis* Miers, Madeira Vine, is native to tropical America, but is cultivated and escaped in southern United States.

There are only about twenty species of Basellaceae, all tropical or subtropical. The family is closely related to the Portulacaceae.

150b Plants not twining or scrambling; ovules 2-many; fruit dehiscent; leaves opposite or alternate .... Purslane Family, PORTULACACEAE

Figure 142

Figure 142 *Claytonia virginica* L., Spring Beauty, is common in the eastern half of the United States; other species occur further west. *Claytonia* is especially notable for the instability in the number of its chromosomes, exemplifying the thought that chromosomes are more nearly comparable to book-cases than to books.

There are about 500 species of Portulacaceae; the principal centers are in western North America and the Andes. Most members of the family are readily recognized by their regular flowers with two sepals, stamens as many as and opposite the petals, free-central placentation, and herbaceous, non-viny habit. See also figs. 88, 175.

151a (from 149b) Ovary with a single style, stigma, and ovule 152

151b Ovary with two or more styles, or with a single but evidently branched style; ovule solitary except in some Caryophyllaceae 153

152a (from 151a) Calyx more or less showy and corolloid, with a well developed tube and lobed limb, sometimes closely subtended by a calyx-like involucre, so that the flower appears to have a calyx and a sympetalous corolla .... Four-O'Clock Family, NYCTAGINACEAE

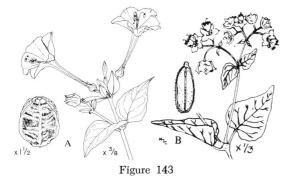

Figure 143

Figure 143   A, *Mirabilis jalapa* L., Four-O'Clock or Marvel of Peru, with a calyx-like involucre beneath a corolla-like calyx; B, *Mirabilis nyctaginea* (Michx.) Macmillan, a wild relative of the Four-O'Clock.

There are about 300 species of Nyctaginaceae, most of them in warm regions. *Abronia*, with rather small but showy flowers in conspicuous involucrate heads, is another familar genus of the Nyctaginaceae, well represented especially in desert regions of western United States. See also figs. 95, 164.

152b Calyx small and inconspicuous, not corolloid; petals none ....
.... Nettle Family, URTICACEAE

Figure 144

Figure 144   *Urtica dioica* L., Stinging Nettle, is a common plant in the Northern Hemisphere.

There are about 700 species of Urticaceae, most of them in warm regions. Some have stinging hairs like the Nettles, but others are harmless. Species of *Pilea* (Aluminum Plant) and *Helxine* (Baby's-tears) are familiar as ornamental foliage plants. *Boehmeria nivea* (L.) Gaud. is the source of the textile fiber ramie.

153a (from 151b) Flowers unisexual; leaves palmately lobed or palmately compound, at least the middle and lower ones opposite ....
.... Hops Family, CANNABACEAE

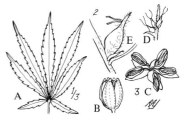

Figure 145

Figure 145 *Cannabis sativa* L., Hemp, is the source of marijuana and hashish as well as of a fiber used in ropes and twine. A, leaf; B, young staminate flower; C, mature staminate flower; D, cluster of pistillate flowers; E, single pistillate flower with its enclosing bract and exserted styles. Long selection for different uses in different parts of the Old World (fiber toward the north, hallucinogen in tropical and subtropical regions) has produced some racial differentiation within the species, but the variation is continuous, and no sterility barriers are known within the genus.

Cannabis, with a single species, and *Humulus* (Hops) with two or three, are the only members of the Cannabaceae, a family closely related to the Moraceae.

153b Flowers perfect or sometimes unisexual, in the latter case the leaves alternate; leaves simple, variously entire or toothed to seldom pinnately lobed **154**

154a (from 153b) Leaves without stipules; flowers with small sepals in a single series, and without petals **155**

154b Leaves with stipules, or if without stipules, then the flowers with an evidently biseriate perianth **156**

155a (from 154a) Perianth dry and more or less scarious or membranous; filaments generally connate below, the androecial tube sometimes even simulating a sympetalous corolla .... **Amaranth Family, AMARANTHACEAE**

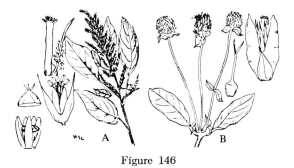

Figure 146

Figure 146 A, *Amaranthus retroflexus* L., Red Root; B, *Gomphrena globosa* L., Globe Amaranth.

The dense inflorescences of *Gomphrena* and *Celosia* (Cockscomb) are showy because the very numerous, tiny flowers have a brightly colored calyx. They are frequently used in dried flower arrangements. Some tropical and subtropical American species of *Amaranthus* were cultivated for grain by the aborigines and have more recently become important in subsistence farming in warm, dry parts of Asia and Africa. There are about 900 species of Amaranthaceae.

155b Perianth more nearly herbaceous, often greenish, only seldom scarious or membranous; filaments distinct, or sometimes connate at the base only .... **Goosefoot Family, CHENOPODIACEAE**

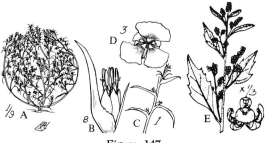

Figure 147

Figure 147 A-D, *Salsola iberica* Sennen & Pau, Russian Thistle, is a common introduced Tumbleweed in western United States. A, habit; B, single flower with subtending bract; C, portion of twig, with axillary flowers; D, fruit, with winged sepals. E, *Chenopodium album* L., Lamb's Quarters, is a common weed of fields and gardens.

There are about 1,500 species of Chenopodiaceae, many of them in dry, alkaline regions. Beets (*Beta vulgaris* L.) and spinach (*Spinacia oleracea* L.) belong to this family. See also fig. 128.

156a (from 154b) Stipules connate to form a well developed, membranous sheath around the stem; stamens most often six .... Buckwheat Family, POLYGONACEAE

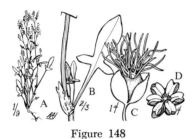

Figure 148

Figure 148 *Rumex acetosella* L., Sourdock, is a common weed in the United States and in Europe. A, habit; B, leaf, with sheathing stipule; C, pistillate flower, with dissected stigmas; D, staminate flower, with six stamens.

Species of *Rumex* have two cycles of three tepals, but all tepals are sepal-like; the members of the outer series often (but not in *R. acetosella*) have a large thickening on the midvein, and the members of the inner series are often closely appressed to the trigonous achene. Some other genera of the Polygonaceae have

more than six stamens and will key elsewhere. See figs. 50, 89.

156b Stipules distinct or connate only at the base, or wanting; stamens most often five, almost never six .... Pink Family, CARYOPHYLLACEAE

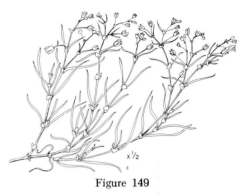

Figure 149

Figure 149 *Spergularia marina* (L.) Grisebach, Sand Spurry, is a species of salt marshes and tidal flats.

Most members of the Caryophyllaceae have ten stamens and will key in another place. See fig. 87.

157a (from 148b) Flowers evidently perigynous, with a well developed hypanthium that may resemble a calyx-tube; ovary variously with 1-several locules 158

157b Flowers hypogynous or nearly so; ovary with two or more locules and axile placentation 161

158a (from 157a) Flowers with a corona in addition to the calyx and corolla, and generally also with a gynophore or androgynophore; plants very often climbing, tendril-bearing vines; ovary unilo-

cular, with 3 (-5) parietal placentas ....
.... Passion-Flower Family,
PASSIFLORACEAE

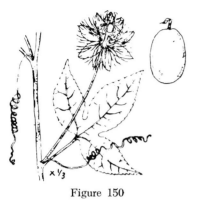

Figure 150

Figure 150 *Passiflora incarnata* L., Passion Flower, is a common, mostly southeastern species with showy, intricately complex flowers. The distinctive floral pattern of *Passiflora* is readily recognizable.

There are about 650 species of Passifloraceae, most of them native to warm regions. More than half the species in the family belong to the genus *Passiflora*.

158b Flowers without a corona, and without a gynophore or androgynophore; plants not climbing vines, lacking tendrils; ovary various **159**

159a (from 158b) Style solitary, with a single (sometimes lobed) stigma; leaves mostly opposite or whorled, seldom alternate; ovary 2-4-locular, with axile placentas ...... Loosestrife Family, LYTHRACEAE

Figure 151

Figure 151 *Lythrum alatum* Pursh, Wing-angled Loosestrife, grows in moist places in the central United States.

Most members of the Lythraceae have about twice as many stamens as petals or sepals, and will therefore key in another place. See figs. 49, 100, 108.

159b Styles two or three and distinct (sometimes bifid), or the stigmas virtually sessile and well separated on the top of the ovary (the ovary sometimes evidently lobed from the summit); leaves mostly alternate; ovary various 160

160a (from 159b) Ovary unilocular, with three parietal placentas; petals red or yellow, seeds with a membranous aril .... Turnera Family, TURNERACEAE

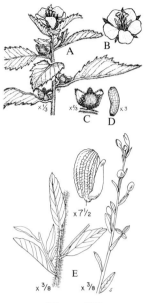

Figure 152

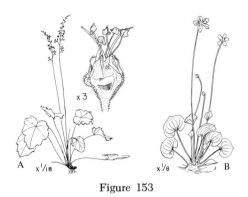

Figure 153

Figure 152   A-D, *Turnera ulmifolia* L., a tropical American species that extends into southern Florida. A, habit; B, flower; C, mature, opened capsule; D, seed. E, *Piriqueta caroliniana* (Walter) Urban, a species of the southeastern coastal plain.

There are about 120 species of Turneraceae, most of them in warm regions; more than half belong to the genus *Turnera*.

160b Ovary bilocular with axile placentas, or unilocular with two or four (not three) parietal placentas; petals white to purple or green; seeds without an aril ......
.... **Saxifrage Family, SAXIFRAGACEAE**

Figure 153   A, *Heuchera americana* L., Alum Root, is widespread in the eastern United States, and several similar species also grow in the West; B, *Parnassia asarifolia* Vent., Grass-of-Parnassus, is another eastern species, but similar ones occur in the West.

There are about 700 species of Saxifragaceae, but many of them have ten stamens and will key elsewhere. The family is here restricted to herbaceous species, but even as so limited it is only loosely knit. Many botanists restrict the family still further and recognize several additional small segregate families, including among others the Parnassiaceae.

161a (from 157b) Leaves evidently lobed or cleft or compound 162

161b Leaves simple and entire or merely toothed 163

162a (from 161a) Style gynobasic, uniting the otherwise separate lobes of the ovary; ovules solitary in each of the two or three locules .... Meadow Foam Family, LIMNANTHACEAE

Figure 154

Figure 154 *Floerkea proserpinacoides* Willd., False Mermaid, a widespread but inconspicuous species. Most members of this small family have eight or ten stamens and will key in another place, but the monotypic genus *Floerkea* generally has only 3-6 stamens. See fig. 113.

162b Style terminal, the lower part tapering-columnar; ovules two in each of the five locules .... Geranium Family, GERANIACEAE

Figure 155

Figure 155 *Erodium cicutarium* (L.) L'Her., Stork's Bill, is an annual weed introduced into America from the Old World. Most members of the Geraniaceae have ten stamens and will key in another place, but *Erodium* has five normal stamens and five staminodes (filaments without anthers). See figs. 114, 161.

163a (from 161b) Plants growing in mud or shallow water; leaves opposite or whorled, with scarious stipules .... Waterwort Family, ELATINACEAE

Members of the Elatinaceae are more or less aquatic, but sometimes they will grow in mud or other wet, low places. See fig. 51.

163b Plants ordinary mesophytes or xerophytes, not growing in very wet places; leaves variously arranged; stipules inconspicuous or wanting 164

164a (from 163b) Seeds with a peripheral embryo curved around the central perisperm; ovules one to more often numerous in each locule; leaves alternate, opposite, or whorled; flowers usually not showy .... Carpet-Weed Family, MOLLUGINACEAE

Figure 156

Figure 156 *Mollugo verticillata* L., Carpet Weed, is a widely distributed, inconspicuous weed that grows flat on the ground.

There are about a hundred species of Molluginaceae, most of them in warm regions.

164b Seeds with a straight embryo, scanty endosperm, and no perisperm; ovules two in each locule; leaves alternate; flowers more or less showy .... Flax Family, LINACEAE

Figure 157

Figure 157 *Linum usitatissimum* L., Flax, has sky-blue flowers. Linen is made from the fibers of its stem, and linseed oil from its seeds.

There are about 220 species in this widespread family, 200 of them in the genus *Linum*. Some species of *Linum* have bright yellow flowers, instead of blue.

165a (from 73b) Ovary with two or more locules **166**

165b Ovary with only one locule **171**

166a (from 165a) Flowers with a conspicuous free spur; herbs **167**

166b Flowers without a free spur; herbs or woody plants **168**

167a (from 166a) Leaves palmately veined, commonly peltate; stamens eight; ovary with three locules, each containing only one ovule; fruit separating into 1-seeded mericarps .... **Nasturtium Family, TROPAEOLACEAE**

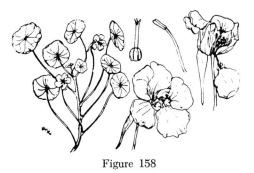

Figure 158

Figure 158 *Tropaeolum majus* L., Garden Nasturtium.

There are about eighty species of Tropaeolaceae, nearly all of them in the genus *Tropaeolum*, which occurs in the mountains from Mexico to Chile. Several species are occasionally grown for ornament. Although very different in other respects, the Tropaeolaceae resemble the Brassicaceae in producing mustard oil.

167b Leaves pinnately veined, not peltate; stamens five; ovary with five locules, each with 3-many ovules; fruit usually an elastically dehiscent capsule that flings out the seeds .... **Touch-Me-Not Family, BALSAMINACEAE**

Figure 159

Figure 159 *Impatiens biflora* Walter, Wild Touch-Me-Not, has orange-yellow flowers with brown spots. The Pale Touch-Me-Not, *I. pallida* Nutt., has pale yellow flowers with brown spots.

Nearly all of the 450 species of Balsaminaceae belong to the genus *Impatiens,* which centers in tropical Asia and Africa. Several species are cultivated as garden ornamentals, notably *I. sultanii* Hook.f., *I. holstii* Engler & Warb., and *I. balsamina* L. The first two of these are closely related and are commonly sold as Impatiens. The third is more distinctive and is called Balsam.

168a (from 166b) Anthers opening by terminal pores; flowers superficially suggesting those of the Fabaceae, but structurally very different; sepals mostly five, the two inner (lateral) ones larger than the others and petaloid; petals usually three, two upper and one lower, the latter boat-shaped and often fringed at the tip; filaments generally connate into a tube that is often joined with the petals toward the base; leaves alternate, in most species simple ....
.... Milkwort Family, POLYGALACEAE

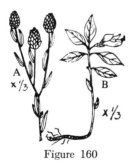

Figure 160

Figure 160 A, *Polygala sanguinea* L., Purple Milkwort; B, *Polygala paucifolia* Willd., Flowering Wintergreen.

About two-thirds of the 750 species of Polygalaceae belong to the genus *Polygala.* Some species are used in folk medicine, and a few are garden flowers.

168b Anthers opening by longitudinal slits; flowers very different from those of the Polygalaceae; leaves mostly opposite, variously simple or compound 169

169a (from 168b) Woody plants; flowers without a spur 170

169b Herbs; flowers with a more or less well developed, nectar-bearing spur adnate to the pedicel ....
...... Geranium Family, GERANIACEAE

Figure 161

Figure 161 *Pelargonium domesticum* Bailey, Lady Washington Geranium, is a cultigen derived from several South African species. Most members of the Geraniaceae have regular flowers and will key elsewhere. See figs. 114, 155.

170a (from 169a) Leaves simple, mostly pinnately veined; fruit indehiscent, but often separating into mericarps; stamens mostly 10, or five plus five sta-

minodes .... **Barbados Cherry Family, MALPIGHIACEAE**

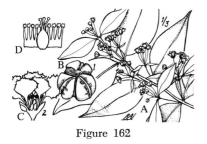

Figure 162

Figure 162 *Malpighia glabra* L., Barbados Cherry, is a tropical shrub with edible fruits that are high in Vitamin C and are used in making jams and preserves. A, habit; B, fruit; C, flower; D, portion of flower, laid open, showing pistil and stamens with basally connate filaments.

There are about 800 species of Malpighiaceae, most of them in the New World tropics. Species of *Banisteriopsis* produce hallucinogens that are used by South American Indians.

170b Leaves palmately compound, the leaflets pinnately veined; fruit a capsule, commonly with a single large seed; stamens (5) 6-8 .... Horse-Chestnut Family, HIPPOCASTANACEAE

Figure 163

Figure 163 *Aesculus glabra* Willd., Ohio Buckeye, is a small tree of moist, rich woods in eastern United States. *Aesculus hippocastanum* L., the Horse-Chestnut, is planted as a street and landscape tree.

There are only about fifteen species of Hippocastanaceae, thirteen of them in the genus *Aesculus*.

171a (from 165b) Placenta only one 172

171b Placentas two to several 176

172a (from 171a) Ovules only one or two; fruit indehiscent; leaves commonly simple, but sometimes compound 173

172b Ovules (2) several or more or less numerous; fruit a legume, usually dehiscent; leaves nearly always compound 175

173a (from 172a) Leaves opposite; herbs .... Four-O'Clock Family, NYCTAGINACEAE

Figure 164

Figure 164 *Wedeliella incarnata* (L.) Cockerell, is a western desert species that can represent the relatively few members of the Nyctaginaceae with irregular flowers. It has three flowers clustered together so as to resemble a single flower. See figs. 95, 143.

173b Leaves alternate, except in a few woody species; plants woody or less often herbaceous **174**

174a (from 173b) Flowers hypogynous; fruit spiny or bristly; stamens four; petals five; sepals mostly five ....
.... **Krameria Family, KRAMERIACEAE**

Figure 165

Figure 165 *Krameria parvifolia* Benth. A, habit, with flowers and spiny fruit; B, detail of part of flower, showing pistil, stamens, and highly irregular corolla.

There are only about twenty species of Krameriaceae, all in the genus *Krameria,* which occurs in arid regions from southwestern United States to Argentina and Chile. *Krameria* is a root parasite, but it has green leaves.

174b Flowers perigynous; fruit not spiny or bristly; stamens four; sepals four; petals wanting ....
.... **Protea Family, PROTEACEAE**

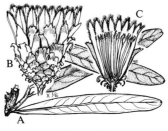

Figure 166

Figure 166 *Protea neriifolia* R. Br. A, leafy twig; B, involucre surrounding a head of flowers; C, long section of flower-head.

There are more than a thousand species of Proteaceae, most of them native to South Africa or Australia. The flower-heads often look remarkably like those of the Asteraceae, but the structure is very different. Macadamia nuts come from the Australian species *Macadamia ternifolia* F. Muell., of this family.

175a (from 172b) Corolla mostly papilionaceous, the upper petal (called the banner or standard) borne externally to the others and generally the largest, folded along the midline so as to enfold the other petals in bud; two lateral petals (called wings) similar to each other and mostly distinct; two lower petals similar to each other and mostly connate distally to form a keel enfolding the stamens and pistil; stamens mostly 10, usually connate by their filaments to form an open or closed sheath around the pistil, the uppermost one often more or less separate from the sheath formed by the other nine, or sometimes the filaments all distinct ....
.... **Pea Family, FABACEAE**

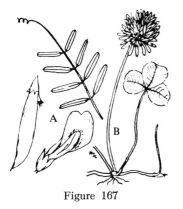

Figure 167

Figure 167   A, *Vicia micrantha* Nutt., Small-flowered Vetch; B, *Trifolium repens* L., White Clover.

There are about 10,000 species of Fabaceae, widespread throughout the world. The distinctive flowers of the family are easily recognizable; only some members of the Caesalpiniaceae have flowers that approach the Fabaceae in structure. Peas, Beans, Alfalfa, Lupine, Wisteria, and Loco Weed (*Astragalus*) belong to the Fabaceae. The Texas Bluebonnet (species of *Lupinus*) is the state flower of Texas; and Red Clover, *Trifolium pratense* L., is the state flower of Vermont.

175b **Corolla not papilionaceous; upper petal usually borne internally to the lateral petals and smaller than them; filaments distinct or variously connate, but not usually forming a definite sheath around the pistil .... Caesalpinia Family, CAESALPINIACEAE**

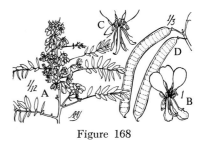

Figure 168

Figure 168   *Cassia marilandica* L., Wild Senna, is common in the southeastern United States. A, habit; B, flower; C, portion of flower, with lower petals removed, showing long and short stamens, long style; D, fruits. There are about 2,200 species of Caesalpiniaceae, only a few in temperate climates. Honey Locust (*Gleditsia triacanthos* L.) and Redbud (*Cercis canadensis* L.) are familiar to us. The Royal Poinciana (*Delonix regia* (Bojer) Raf.) is a common ornamental tree in the tropics. Many botanists treat the Caesalpiniaceae and Mimosaceae as subfamilies of the family Fabaceae. The family as thus broadly defined is often called Leguminosae.

176a **(from 171b) Sepals two; stamens mostly six and connate into two lateral phalanges; petals four, in two sets of two; herbs ....**
...... Fumitory Family, FUMARIACEAE

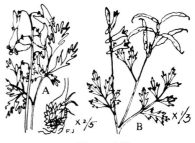

Figure 169

Figure 169 A, *Dicentra cucullaria* (L.) Bernh., Dutchman's Breeches; B, *Corydalis aurea* Willd., Golden Corydalis. Bleeding Heart (*Dicentra spectabilis* (Don.) Lemaire) is another familiar member of the Fumariaceae.

There are about 400 species in this family, most in North Temperate regions. The combination of two sepals, a superior ovary, and an irregular corolla immediately sets the Fumariaceae apart from nearly all other dicotyledons.

176b Sepals four or five (-8); stamens of varying number, but not connate into two lateral phalanges; petals 4-8; herbs or woody plants **177**

177a (from 176b) Ovary open at the top, the (2) 3-6 (7) stigmas sessile around its rim; sepals 4-8; petals 4-8; stamens 3-40; mostly herbs .... Mignonette Family, **RESEDACEAE**

Figure 170

Figure 170 *Reseda lutea* L., Yellow Cut-leaved Mignonette, is a European species that has gone wild in the eastern United States. A related species, *R. odorata* L., with mostly undivided leaves, is the Garden Mignonette.

There are about seventy species of Resedaceae, all in the Northern Hemisphere, only a

few in the United States. The open ovary, with exposed ovules, is a paradox among angiosperms, which otherwise have the ovules enclosed at the time of flowering.

177b Ovary closed; style solitary, with a simple or shortly lobed stigma, or the stigma sessile atop the ovary; sepals, petals, and stamens of varying number; herbs or woody plants **178**

178a (from 177b) Sepals, petals, and stamens generally each five; placentas mostly three; fruit without a replum **179**

178b Sepals and petals mostly each four; stamens 6-many; placentas mostly two; fruit generally with a replum (see key lead number 146 for definition) .... Caper Family, **CAPPARACEAE**

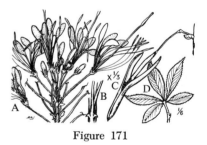

Figure 171

Figure 171 *Cleome spinosa* L., Spider Flower, is a garden ornamental from tropical America that sometimes grows wild in the United States. A, top of inflorescence; B, spiny leaf-base; C, fruit undergoing dehiscence, showing valves and replum; D, leaf.

Most species of Capparaceae have regular flowers and will key elsewhere. See figs. 82, 140.

179a (from 178a) Flowers distinctly perigynous; woody plants with compound leaves; five stamens alternating with as many evident staminodes .... Horse-Radish Tree Family, MORINGACEAE

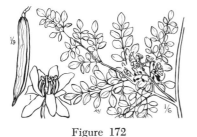

Figure 172

Figure 172 *Moringa oleifera* Lam., Horse-Radish Tree, a native of India, is cultivated in Florida and in tropical regions generally. Its seeds are the source of Ben Oil, used in perfumery and lubrication.

There are only about ten species of Moringaceae, all in the tropical genus *Moringa*

179b Flowers hypogynous; our spp. herbs; leaves simple to sometimes dissected, but without well defined leaflets; no staminodes .... Violet Family, VIOLACEAE

Figure 173

Figure 173 A, *Viola pedata* L., Bird's-foot Violet; B, *Viola striata* L., Pale Violet.

There are about 800 species of Violaceae, half of them in the genus *Viola*, which is common in North Temperate regions. Several states have chosen violets as the state flower. Violet flowers are easily recognized by their aspect, having a large lower petal with a basal spur, two smaller lateral petals, and two upper petals more or less similar to the lateral ones. Some of the tropical members of the family, on the other hand, have flowers of much more ordinary appearance.

180a (from 72b) Plants distinctly succulent; flowers perfect 181

180b Plants not succulent, or if somewhat succulent, then the flowers unisexual 183

181a (from 180a) Plants either stem-succulents, or distinctly spiny, or usually both, only rarely with well developed leaves; ovary unilocular .... Cactus Family, CACTACEAE

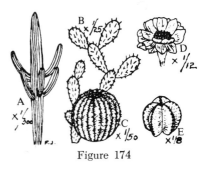

Figure 174

Figure 174 A, *Carnegiea gigantea* (Engelm.) Britt. & Rose, Giant Saguaro, is the state flower of Arizona and is the largest known cactus; it may reach a height of 15 m; B, *Opuntia engelmannii* Salm-Dyck, Prickly Pear; C, *Echinocactus grusonii* Hild., Golden Cactus; E, *Astrophytum myriostigma* Lemaire, Bishop's Cap; D, typical cactus flower.

There are about 2,000 species of Cactaceae, nearly all in the New World. Most of them are desert plants, but a few grow in more ordinary, mesic habitats, and some are epiphytes in tropical forests. The only other plants that closely resemble the spiny cacti in vegetative aspect are some of the African species of *Euphorbia;* these have milky juice (which cacti do not), and have very different flowers. See also fig. 198.

181b Plants leaf-succulents, not spiny 182

182a (from 181b) Ovary unilocular; fruit a circumscissile capsule; petals often broad ....
Purslane Family, PORTULACACEAE

Figure 175

Figure 175 A, *Portulaca oleracea* L., Purslane; B, *Portulaca grandiflora* Hook., Rose Moss.

Although most members of the Portulacaceae have a superior ovary, *Portulaca* has the ovary half-inferior and might be sought here in the key. See figs. 88, 142.

182b Ovary with several locules; fruit most commonly a loculicidal capsule, in any case not circumscissile; petals usually slender ....
...... Fig Marigold Family, AIZOACEAE

See fig. 98. The ovary in plants in the family Aizoaceae ranges from superior through partly inferior to wholly inferior; thus many members of the family will key in another place.

183a (from 180b) Ovary and fruit 6-locular; stamens six or twelve; leaves palmately veined, in most species cordate, reniform, or hastate; plants very often climb-

ing vines or stemless herbs, seldom leafy-stemmed erect herbs .... Birthwort Family, **ARISTOLOCHIACEAE**

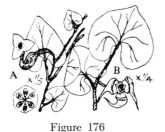

Figure 176

Figure 176   A, *Aristolochia tomentosa* Sims, Woolly Pipe-Vine; B, *Asarum canadense* L., Wild Ginger.

About 500 of the 600 species of Aristolochiaceae are woody climbers and belong to the genus *Aristolochia. Asarum* and *Hexastylis*, on the other hand, are herbs with leaves and flowers at the ground-level. See also fig. 110.

183b Plants not with the above combination of characters, usually differing in more than one respect 184

184a (from 183b) Stamens mostly numerous, rarely as few as eight or ten 185

184b Stamens relatively few, up to about ten 191

185a (from 184a) Plants herbs or occasionally soft shrubs 186

185b Plants shrubs or trees 187

186a (from 185a) Flowers perfect; ovary mostly with a single locule and parietal placentas; sepals mostly five; style one; leaves without stipules .... Loasa Family, **LOASACEAE**

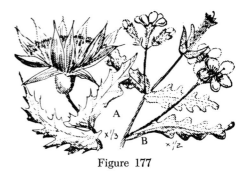

Figure 177

Figure 177   A, *Mentzelia decapetala* (Pursh) Urban & Gilg, Prairie Lily; B, *Mentzelia albicaulis* Dougl., White-stemmed Mentzelia.

There are about 200 species of Loasaceae, nearly all of them in the New World. *Mentzelia* is the largest genus of the family in the United States. Members of the Loasaceae are usually very rough-hairy, the hairs often silicified, sometimes stinging.

186b Flowers unisexual; ovary mostly with three locules and axile placentas; sepals two (-5); styles distinct, except sometimes at the base, usually three, sometimes bifid; leaves without stipules .... Begonia Family, **BEGONIACEAE**

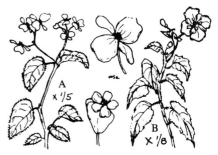

Figure 178

Figure 178   A, *Begonia* x *semperflorens-cultorum* Hort.; B, *Begonia* x *tuberhybrida* Voss, Tuberous Begonia.

There are about a thousand species of Begoniaceae, most of them in the large genus *Begonia*. Northern South America is their greatest center. Many species are cultivated for ornament, and numerous hybrids have been produced.

187a (from 185b) Leaves opposite **188**

187b Leaves alternate **190**

188a (from 187a) Leaves with scattered, embedded secretory cavities that appear as translucent dots when the leaf is held up to the light; style solitary, with a capitate stigma; ovary usually with 3-5 locules ....
.... **Myrtle Family, MYRTACEAE**

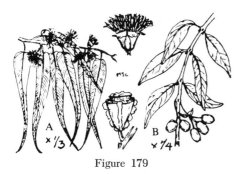

Figure 179

Figure 179   A, *Eucalyptus globulus* Labill., Tasmanian Blue Gum, is often planted in California; B, *Syzygium paniculatum* Gaertn., Australian Brush-Cherry, is also cultivated in California.

Various species of *Eucalyptus* are important timber trees in southern Australia, especially in the strip along the coast that has a Medi-terranean climate (hot, dry summers and mild, moist winters). Much of California has a similar climate, and *Eucalyptus* does well there. Guava, *Psidium guajava* L., is another member of this large, chiefly tropical and subtropical family, which has about 3,000 species.

188b Leaves without evident secretory cavities; other features various **189**

189a (from 188b) Style solitary, with a capitate stigma; ovary commonly with 7-9 carpels, these often superposed in two layers; fruit large, fleshy, with a leathery rind ....
.... **Pomegranate Family, PUNICACEAE**

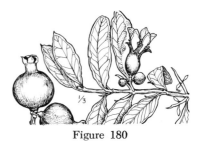

Figure 180

Figure 180   *Punica granatum* L., Pomegranate, is a cultigen that also grows wild from the Balkan region to northern India.

The family has only one other species, *Punica protopunica* Balf.f., on the island of Socotra.

189b Styles usually several, commonly 3-5, sometimes joined at the base, only rarely joined throughout; ovary with (2) 3-5 (-12) carpels; fruit usually a capsule, seldom a small berry .... **Hydrangea Family, HYDRANGEACEAE**

Figure 181

Figure 181 *Philadelphus coronarius* L., Mock-Orange, also called Syringa.

There are about 250 species of Hydrangeaceae, most of them in three genera, *Hydrangea*, *Philadelphus*, and *Deutzia*. All three genera are cultivated as ornamentals. *Philadelphus lewisii* Pursh (named for the explorer Meriwether Lewis) is the state flower of Idaho.

**190a (from 187b) Styles 2-several; leaves with stipules; fruit a pome ....
.... Rose Family, ROSACEAE**

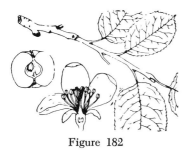

Figure 182

Figure 182 *Pyrus malus* L., Apple.

Only one subfamly of Rosaceae, the Pomoideae, will key here. This group includes Pears, Quince, Hawthorn (*Crataegus*), Mountain Ash (*Sorbus*) and Service Berry (*Amelanchier*) as well as apples. Hawthorn is the state flower of Missouri. Other members of the Rosaceae have the ovary or ovaries superior and will key elsewhere. See figs. 27, 42, 92.

**190b Style one; leaves without stipules; fruit a capsule, or less often a berry or drupe
.... Brazil Nut Family, LECYTHIDACEAE**

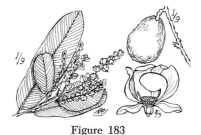

Figure 183

Figure 183 *Bertholletia excelsa* Humb. & Bonpl., Brazil Nut, an Amazonian forest tree.

There are about 450 species of Lecythidaceae, many of them in the rain forests of tropical South America. None is native to the United States.

191a (from 184b) Ovules and seeds several or numerous in each of the 2-several locules of the ovary; herbs or shrubs. (Some species of *Heuchera*, in the Saxifragaceae, have the ovary virtually inferior, and would be sought here except that the ovary is unilocular) 192

191b Ovules and seeds 1-3 in the single locule, or 1-2 in each of the 2-several locules of the ovary 195

192a (from 191a) Anthers opening by terminal pores; leaves with 3-9 prominent, subparallel longitudinal veins ...... Melastome Family, MELASTOMATACEAE

Figure 184

Figure 184 *Rhexia virginica* L., Meadow-Beauty, is one of the few species of Melastomataceae in the United States.

Most of the 4,000 species of the family are tropical, many of them in South America. The family is easily recognized by its leaf venation, epigynous flowers, and unusual anthers, which often have appendages of one or another sort.

192b Anthers opening by longitudinal slits; leaves pinnately or palmately veined **193**

193a (from 192b) Shrubs; flowers usually perfect; ovary most commonly 2-3-locular; leaves alternate .... Gooseberry Family, GROSSULARIACEAE

Figure 185

Figure 185 *Ribes missouriense* Nutt., Missouri Gooseberry.

There are more than 300 species of Grossulariaceae, nearly half of them in the genus *Ribes*, which includes Gooseberries (usually spiny) and Currants (unarmed). *Ribes* is the alternate host for the fungus that causes White Pine Blister Rust.

193b Herbs (some spp. of *Fuchsia*, in the Onagraceae, are somewhat shrubby, but have a 4-locular ovary and may have opposite leaves) **194**

194a (from 193b) Styles distinct, as many as the carpels, sometimes bifid; petals wanting or vestigial; flowers unisexual, or staminate on some plants and perfect on others; ovary unilocular, with several parietal placentas; leaves deeply pinnatifid to pinnately compound .... Datisca Family, DATISCACEAE

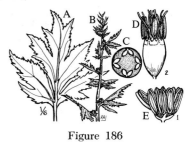

Figure 186

Figure 186 *Datisca glomerata* (Presl) Baill., a species native to California. A, leaf; B, upper part of plant; C, diagrammatic cross-section of ovary; D, perfect flower; E, staminate flower.

There are only four species of Datiscaceae, two in Indo-Malaysia, one on the Asian mainland, and one in western United States.

194b Style solitary, with a capitate to 4-lobed stigma; flowers perfect, usually with evident petals; ovary most commonly (but not always) 4-locular; leaves simple and entire to more or less lobed or dissected .... **Evening Primrose Family, ONAGRACEAE**

Figure 187

Figure 187 *Oenothera biennis* L., Evening Primrose. Flowers of this and other species of *Oenothera* open in the evening and wilt the next day.

There are about 650 species of Onagraceae, many of them in western United States. Species of *Fuchsia* and *Clarkia*, as well as *Oenothera* are cultivated for ornament. Most members of the family stand apart from other angiosperms in their combination of 4-merous flowers, inferior ovary, and four or eight stamens, which are of ordinary type instead of specialized as in the Melastomataceae.

195a (from 191b) **Leaves mostly compound or dissected (seldom simple, as in *Hedera*, of the Araliaceae, a climbing, woody vine with palmately lobed leaves, or in a few members of the Apiaceae, which have distinctive, schizocarpic fruits)** 196

195b Leaves simple, not compound or dissected; fruit not schizocarpic; plants only seldom climbing; inflorescence not umbellate 197

196a (from 195a) **Carpels 1-many, most often five; fruit usually a drupe or a berry, only rarely a schizocarp with a carpophore; flowers commonly in umbels or heads that are grouped into various sorts of compound inflorescences, but only rarely forming regular, compound umbels; trees, shrubs, or woody vines, only a few species (including some of our native ones) herbaceous** .... **Ginseng Family, ARALIACEAE**

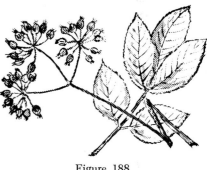

Figure 188

Figure 188 *Aralia nudicaulis* L., Wild Sarsaparilla.

Most of the 700 species of Araliaceae are woody plants of warm regions, but Ginseng (*Panax*) is a herb of forests in eastern United States and in China, and English Ivy (*Hedera helix* L.) is another familiar temperate-zone member of the family. *Schefflera* and *Polyscias*, both tropical, are often cultivated indoors as foliage plants.

196b Carpels consistently two; fruit a dry schizocarp, the mericarps usually attached to a persistent central carpo-

phore; flowers commonly in compound umbels, less often in simple umbels or other sorts of inflorescences; herbs .... Carrot Family, APIACEAE

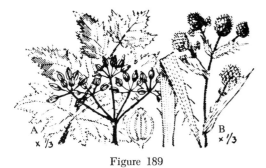

Figure 189

Figure 189 A, *Pastinaca sativa* L., Parsnip; B, *Eryngium yuccifolium* Michx., Rattlesnake-Master, is unusual in its simple leaves and capitate inflorescences, but its flowers and fruits are typical of the Apiaceae.

Among the 3,000 species of Apiaceae are several food plants (Carrot, Parsley, Celery), many plants used in seasoning (Anise, Caraway, Coriander, Cumin, Dill, Lovage), and some highly poisonous plants (Water Hemlock, Poison Hemlock). The compound umbels immediately set most members of the Apiaceae apart from virtually all other angiosperms, and the fruits are also distinctive. The plants characteristically have well developed internal secretory canals that contain various sorts of aromatic substances. The word aromatic is here used in a nontechnical sense.

197a (from 195b) Ovary unilocular; herbs or woody plants 198

197b Ovary with two or more locules; woody plants, except *Cornus canadensis* 201

198a (from 197a) Herbs with the foliage leaves all basal; perianth dimerous or none 199

198b Woody plants, or herbs with the leaves distributed along the stem; perianth not dimerous 200

199a (from 198a) Inflorescence a large panicle; leaves often very large; stamens one or two; sepals and petals each two, or the petals (and sometimes also the sepals) wanting .... Gunnera Family, GUNNERACEAE

Figure 190

Figure 190 *Gunnera chilensis* Lam. is sometimes grown as a curiosity because of its spectacularly large leaves, nearly a meter across. A, habit; B, male flower; C, female flower; D, portion of inflorescence.

The family has only the genus *Gunnera*, with about forty species, all tropical or in the Southern Hemisphere. The Gunneraceae are usually considered to be related to the chiefly aquatic family Haloragaceae.

199b Inflorescence a dense spike subtended by an involucre of petal-like bracts; stamens 6-8; perianth none ...... Lizard's-Tail Family, SAURURACEAE

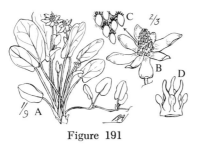

Figure 191

Figure 191 *Anemopsis californica* Hook., Yerba Mansa, grows in alkaline, marshy places from California to Texas. A, habit; B, inflorescence, with large, petal-like bracts at the base; C, portion of inflorescence, with reflexed bracts beneath the individual flowers; D, flower. See fig. 38 for further comments about the Saururaceae.

200a (from 198b) **Perianth strictly monochlamydeous, of a single set of (3) 4-5 (6) tepals that may be joined into a tube toward the base; stamens as many as the tepals; ovules pendulous from the tip of the free-central placenta; root parasites .... Sandlewood Family, SANTALACEAE**

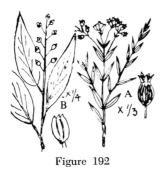

Figure 192

Figure 192 A, *Comandra umbellata* (L.) Nutt., False Toadflax; B, *Pyrularia pubera* Michx., Oil-Nut.

There are about 400 species of Santalaceae, some herbs, as *Comandra*, some shrubs, as *Pyrularia*, some trees, such as the famed Sandalwood (*Santalum*) of the Far East. Members of the Santalaceae are root-parasites that also have green leaves and make much of their own food.

200b **Perianth monochlamydeous (as in *Terminalia*) or often dichlamydeous, of 4-5 (8) sepals or calyx-lobes and often also as many petals; stamens usually twice as many as the sepals; ovules pendulous from the tip of the locule; woody plants, not parasitic, sometimes of mangrove habit .... Indian Almond Family, COMBRETACEAE**

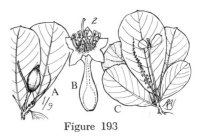

Figure 193

Figure 193 *Terminalia catappa* L., Indian Almond, is widely cultivated in the tropics as a shade tree and for its edible seeds and tanniferous bark. A, leafy twig with a single fruit; B, a single perfect flower, with the stamens attached at the summit of the hypanthium; C, leafy twig with flowers. The lower flowers of the spike are perfect, and the upper ones are staminate.

There are about 500 species of Combretaceae, all tropical and subtropical, most of them in the two large genera *Combretum* and *Terminalia*.

201a (from 197b) Stamens eight or ten **202**

201b Stamens four or five **203**

202a (from 201a) Leaves opposite; fruit a berry; some species mangroves ....
.... Red Mangrove Family, **RHIZOPHORACEAE**

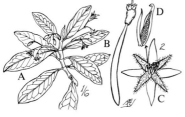

Figure 194

Figure 194 *Rhizophora mangle* L., American Mangrove, is the most familiar species of Rhizophoraceae in the New World. It forms dense thickets in shallow water along tropical coasts. A, flowering twig; B, germinating seedling, with elongate radicle; C, flower; D, stamen.

Only about twenty of the hundred species of Rhizophoraceae are mangroves, but the others do not attract so much attention. The family is wholly tropical.

202b Leaves alternate; fruit a drupe; plants not mangroves .... Sour Gum Family, **NYSSACEAE**

Figure 195

Figure 195 *Nyssa sylvatica* Marsh., Black Gum or Sour Gum, is a common tree in eastern United States. Its leaves turn bright red in the fall.

There are only seven or eight species of Nyssaceae, making up three genera, *Nyssa*, *Davidia*, and *Camptotheca*.

203a (from 201b) Fruit a drupe; leaves opposite, only rarely alternate ....
.... Dogwood Family, **CORNACEAE**

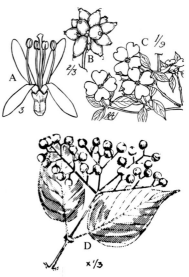

Figure 196

Figure 196 A-C, *Cornus florida* L., Flowering Dogwood, the state flower of North Carolina and Virginia, has four large, white (seldom pink), petal-like bracts around the small, tight cluster of flowers; A, single flower; B, cluster of fruits; C, flowering branch. D, *Cornus drummondii* C. A. Meyer, Roughleaf Cornel, like many other species of *Cornus*, has a more open inflorescence, without the large bracts.

There are about a hundred species of Cornaceae, half of them in the genus *Cornus*. *Cornus alternifolia* L. is unusual in its alternate leaves, but otherwise it looks much like several other shrubby species of the genus.

203b Fruit a capsule; leaves alternate .... **Witch Hazel Family, HAMAMELIDACEAE**

Figure 197

Figure 197 *Hamamelis virginiana* L., Witch Hazel, blooms in late autumn, after the leaves have fallen. Some other species of *Hamamelis* bloom in late winter or earliest spring. The flowers of *Hamamelis* appear to be adapted to pollination by insects, yet they bloom when few insects are about, at a time when trees are bare of leaves and wind-pollination is more expectable.

There are about a hundred species in the Hamamelidaceae; thirteen of the twenty-six genera are monotypic. The Hamamelidaceae play a key role in much phylogenetic speculation about the evolutionary origin of such wind-pollinated, catkin-bearing families as the Betulaceae and Fagaceae. See also fig. 62.

204a (from 46b) Functional stamens either more numerous than the corolla-lobes, or as many as the corolla-lobes and opposite them **205**

204b Functional stamens as many as the corolla lobes and alternate with them, or fewer than the corolla lobes **218**

205a (from 204a) Plants succulent, leafless, often spiny; corolla lobes more or less numerous, sometimes in more than one cycle ...... **Cactus Family, CACTACEAE**

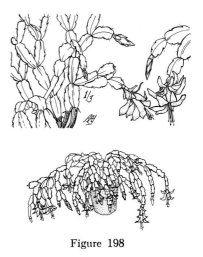

Figure 198

Figure 198 *Schlumbergera bridgesii* (Lem.) Loefgr., Christmas Cactus, is one of several spineless cacti that are adapted to growing as epiphytes in tropical forests. The petals in this and some other cacti are borne in several cycles and all joined together toward the base to form a tube. See also fig. 174.

205b Plants otherwise, only seldom succulent, always with evident leaves for at least a part of the season; corolla-lobes ordinarily in a single cycle, seldom more than six in all (double-flowered specimens of *Rhododendron* have ten corolla-lobes in two series) 206

206a (from 205b) Placentation free-central or basal in a unilocular ovary; functional stamens as many as and opposite the corolla-lobes, typically five 207

206b Placentation axile or parietal or marginal; functional stamens as many as or very often more numerous than the corolla-lobes 210

207a (from 206a) Herbs or sometimes low shrubs; fruit dry, either dehiscent or indehiscent 208

207b Shrubs or trees; fruit mostly fleshy and indehiscent; ovules several or numerous on a free-central placenta 209

208a (from 207a) Ovule solitary and basal; style one, or styles five; fruit most commonly an achene, but sometimes a capsule .... Leadwort Family, **PLUMBAGINACEAE**

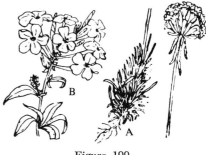

Figure 199

Figure 199 A, *Armeria maritima* Willd., Sea Pink, is a native of European sea coasts that is often cultivated for ornament; B, *Plumbago capensis* Thunb., Leadwort, is a South African species, cultivated in the southern United States.

There are about 400 species of Plumbaginaceae. The family is taxonomically isolated, and its relationships are uncertain. It may or may not be allied to the Primulaceae.

208b Ovules several or numerous on a free-central placenta; style one; fruit generally a capsule .... Primrose Family, **PRIMULACEAE**

Figure 200

Figure 200 A, *Dodecatheon meadia* L., Shooting Star, is an eastern American species; several other species of *Dodecatheon* grow in the western United States; B, *Lysimachia ciliata* L., Fringed Loosestrife, grows wild in much of the United States.

There are about 800 species of Primulaceae, most of them in temperate regions. The family is easily recognized by its herbaceous habit, free-central placentation, and sympetalous flowers with the stamens attached to the corolla-tube opposite the lobes.

209a (from 207b) Flowers with evident staminodes alternating with the corolla-lobes; plants without an evident secretory system, the leaves not gland-dotted .... **Theophrasta Family, THEOPHRASTACEAE**

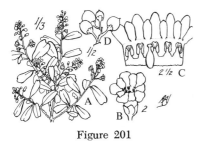

Figure 201

Figure 201 *Jacquinia keyensis* Mez, Joeweed, of southern Florida, is the only member of this tropical family to reach the United States. A, habit; B, flower; C, flower, laid open; D, fruits. *Jacquinia* has relatively large, petaloid staminodes.

There are about a hundred species of Theophrastaceae, half of them in the genus *Jacquinia*.

209b Flowers without staminodes; plants with a well developed secretory system in the stem and leaves, the leaves gland-dotted .... **Myrsine Family, MYRSINACEAE**

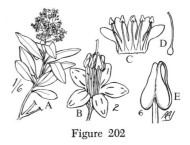

Figure 202

Figure 202 *Ardisia escallonioides* Schlecht. & Cham., Marlberry, is a large shrub or small tree of the Caribbean region, including southern Florida. It has small, white flowers and black berries. A, habit; B, flower, showing glandular-streaked petals; C, corolla, laid open to show attachment of stamens; D, pistil; E, anther.

There are about a thousand species of Myrsinaceae, most of them in warm regions, few of any economic importance.

210a (from 206b) Pistil composed of a single carpel, with a single locule, a single marginal placenta, and a single style and stigma; leaves usually pinnately or bipinnately compound; fruit a legume, generally dry and dehiscing along two sutures .... **Mimosa Family, MIMOSACEAE**

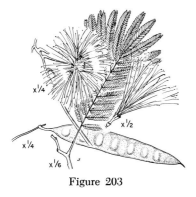

Figure 203

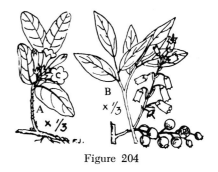

Figure 204

Figure 203 *Albizia julibrissin* (Willd.) Durazz., Silk Tree, or Mimosa, is native to Asia, but is often grown as a landscape tree in the southern United States. Further north it may thrive through mild years but die back or be killed by a severe winter.

There are about 2,000 species of Mimosaceae, most of them in warm (often dry) climates. See also fig. 90. Not many species outside the three families of legumes (Fabaceae, Caesalpiniaceae, and Mimosaceae) have compound leaves and flowers with a single pistil composed of a single carpel.

210b Pistil composed of more than one carpel, as shown by the presence of more than one locule or placenta, and sometimes also by the presence of more than one style or stigma; leaves simple or sometimes palmately or ternately compound; fruit various, but not as above 211

211a (from 210b) Anthers of unusual structure, generally opening by pores or short slits at the apparent tip, often with conspicuous appendages; filaments free from the corolla; style one, unbranched .... Heath Family, ERICACEAE

Figure 204 A, *Epigaea repens* L., Trailing Arbutus; B, *Vaccinium corymbosum* L., Tall Blueberry.

There are about 2,500 species of Ericaceae, many of them grown for ornament, some for food (Blueberry, Cranberry, Huckleberry, Lingonberry). Members of the Ericaceae are strongly mycotrophic (i.e., they have a symbiotic association with fungi that inhabit their roots and also extend out into the soil). In the related, more specialized family Monotropaceae, the plant lacks chlorophyll and depends on the fungus for its food. *Rhododendron* (including Azalea), *Kalmia* (Mountain Laurel), *Pieris* (Andromeda), *Erica* (Heath), and *Calluna* are some familiar ornamentals in the Ericaceae. Mountain Laurel (*Kalmia latifolia* L.) is the state flower of Connecticut and Pennsylvania, and different species of *Rhododendron* are the state flowers of Washington and West Virginia. See also fig. 215.

211b Anthers of ordinary structure, generally opening by longitudinal slits and without conspicuous appendages; filaments free from or often attached to the corolla tube; styles various, often branched or more than one 212

212a (from 211b) Stamens paired beneath the sinuses of the corolla, each with only a single pollen sac; delicate herb with ternately compound leaves .... Adoxa Family, ADOXACEAE

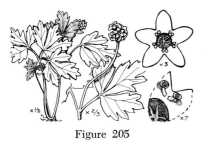

Figure 205

Figure 205 *Adoxa moschatellina* L., Moschatel, is the only member of its family. It is a circumboreal species, extending south to parts of the northern United States.

212b Stamens not so arranged, and generally each with two pollen sacs; woody plants 213

213a (from 212b) Style evidently branched, or the styles more than one 214

213b Style solitary, unbranched, with a simple or merely lobed stigma 216

214a (from 213a) Flowers perfect; thorny, xerophytic shrubs with small, ephemeral leaves; style 3-branched; fruit a capsule .. Ocotillo Family, FOUQUIERIACEAE

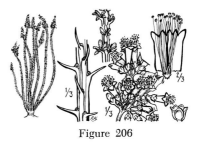

Figure 206

Figure 206 *Fouquieria splendens* Engelm., Ocotillo, is native to deserts of southwestern United States and northern Mexico; its wand-like stems have pyramidal clusters of flame-red flowers in the spring. Its leaves are small and drop off as soon as the soil dries out.

There are only eleven species of Fouquieriaceae, all American desert plants.

214b Flowers mostly unisexual; mostly un-armed shrubs or small trees with well developed leaves; style-branches or styles usually more than three 215

215a (from 214b) Trunk branched; leaves simple and entire, not notably large; styles or style branches most often four; fruit sometimes fleshy, but not melon-like .... Ebony Family, EBENACEAE

Figure 207

Figure 207 *Diospyros virginiana* L., Common Persimmon. Fruits of Persimmon are very bitter at first, but tasty when fully ripe, especially after frost.

Most of the 450 species of this chiefly tropical family belong to the genus *Diospyros*. *Diospyros ebeneum* Koenig, of tropical Asia and the East Indies, is the source of ebony wood.

215b Trunk generally unbranched and with a terminal cluster of large, palmately lobed or palmately compound leaves; styles five; fruit large, fleshy, melon-like .... Papaya Family, CARICACEAE

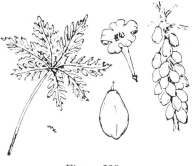

Figure 208

Figure 208 *Carica papaya* L., Papaya, is a tropical American tree with large, delicious fruits. The latex is the source of an enzymatic meat-tenderizer, papain.

There are only about forty-five species of Caricaceae, nearly all in the New World, mostly in tropical and subtropical climates.

216a (from 213b) Plants with milky juice; pubescence generally of 2-armed (pick-shaped) hairs .... Sapodilla Family, SAPOTACEAE

Figure 209

Figure 209 *Bumelia lycioides* (L.) Pers., Buckthorn Bumelia, is native to southwestern United States, but most of the 800 species in the family are tropical.

Manilkara zapota (L.) van Royen, a native of the Caribbean region, is the traditional source of chicle for chewing gum. *Chrysophyllum cainito* L., the Star Apple, is another well known tropical American member of the family. It has apple-sized but soft, edible fruits with the large, flat seeds spreading out in star-like fashion from the center of the fruit, as viewed in cross-section. As in several other species of *Chrysophyllum,* the lower surface of the leaves is densely covered with appressed, shiny, rufous-red hairs. Many species of Sapotaceae have edible fleshy fruits.

216b Plants without milky juice; pubescence not of 2-armed hairs 217

217a (from 216b) Pubescence of stellate hairs (each hair with several branches from the base), varying to peltate scales; stamens all in a single series, not fascicled; anthers narrow, linear or nearly so; fruit mostly dry, seldom fleshy .... Storax Family, STYRACACEAE

Figure 210

Figure 210 *Halesia carolina* L., Silver-bell Tree, is native to southeastern United States.

There are about 150 species of Styracaceae, most of them in the genus *Styrax*. Some species yield gummy resins, such as storax and gum benzoin, used in perfumery.

217b Pubescence neither of stellate hairs nor of peltate scales, sometimes wanting; stamens in more than one series, or grouped into fascicles; anthers broadly ovate to rotund; fruit more or less fleshy .... Symplocos Family, SYMPLOCACEAE

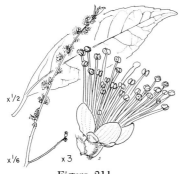

Figure 211

Figure 211 *Symplocos tinctoria* (L.) L'Herit., Sweet-Leaf, is native to southeastern United States. A yellow dye is extracted from the leaves and bark of this and some related species.

The family has only the one genus, with three or four hundred species, most of them in moist, tropical regions. Many members of the Symplocaceae accumulate aluminum in distinctive intercellular bodies.

218a (from 204b) Ovary superior 219

218b Ovary inferior, or sometimes only half-inferior 242

219a (from 218a) Pistil of a special type, consisting of two carpels (as often shown by a 2-lobed stigma), but each carpel divided in half by a median partition, so that the ovary has four chambers, each with a single ovule; ovary usually 4-lobed, from slightly to conspicuously so; fruit separating at maturity into four 1-seeded nutlets (sometimes one or more of the nutlets failing to mature) 220

219b Pistil of a more ordinary type, the number of chambers no greater than the number of carpels, each chamber with 1-many ovules; ovary not 4-lobed; fruit not separating into 1-seeded nutlets .. 222

220a (from 219a) Leaves mostly alternate, usually entire; flowers mostly regular or nearly so and with as many stamens as corolla lobes (typically five of each); plants not aromatic; ovary and style varying from as in the Verbenaceae to much more often as in the Lamiaceae; stems not square .... Borage Family, BORAGINACEAE

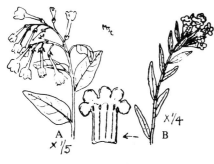

Figure 212

Figure 212   A, *Mertensia virginica* (L.) Pers., Virginia Bluebell; B, *Lithospermum canescens* (Michx.) Lehm., Hoary Puccoon.

There are about 2,000 species of Boraginaceae. *Heliotropium, Cynoglossum* (Hound's-Tongue), and *Myosotis sylvatica* Hoffm. (Forget-Me-Not) are some familiar members of the family. The Forget-Me-Not is the state flower of Alaska. Differences in the structure and ornamentation of the nutlets frequently mark genera and species in the Boraginaceae, but some are more variable than others. *Echium vulgare* L., Blue-weed, with showy blue flowers, is unusual in the family in having a distinctly irregular corolla; a native of Europe, it is well established as a weed in northeastern United States. Many members of the Boraginaceae are notably rough-hairy, the hairs with silicified or calcified walls.

220b **Leaves opposite or (sometimes whorled), entire or often toothed or cleft or even compound; flowers with a more or less irregular corolla and two or four stamens (some exceptions in the Verbenaceae); young stems commonly square** **221**

221a (from 220b) Style terminal or nearly so, the summit of the ovary only shortly or not at all impressed-lobed; plants seldom aromatic .... **Verbena Family, VERBENACEAE**

Figure 213

Figure 213   A, *Verbena canadensis* (L.) Britton, Large-flowered Verbena; B, *Phyla cuneifolia* (Torr.) Greene; C, *Callicarpa americana* L., American Beautyberry, a shrub with pink to purple or blue fruits.

There are about 2,000 species of Verbenaceae, most of them tropical. *Lantana* is a familiar ornamental shrub in this family. *Tectona grandis* L.f., originally of India and Burma, is the source of teak wood. Although the Verbenaceae and Lamiaceae have traditionally been held as distinct families, the two groups are confluent, and the taxonomic boundary between them is arbitrary. See also fig. 233.

221b Style commonly gynobasic, uniting the four otherwise essentially distinct lobes of the ovary, or less often the ovary lobed only part way (one-third or more) from the summit toward the base; plants commonly aromatic .... **Mint Family, LAMIACEAE**

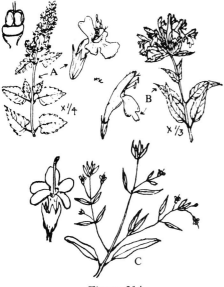

Figure 214

Figure 214 A, *Nepeta cataria* L., Catnip; B, *Monarda fistulosa* L., Horse Mint; C, *Isanthus brachiatus* (L.) B.S.P., False Pennyroyal.

There are more than 3,000 species of Lamiaceae. *Coleus*, *Lavandula* (Lavender), *Marrubium* (Horehound), *Mentha* (Mint), *Salvia* (Sage), and *Thymus* (Thyme) are some of the more familiar genera. *Salvia* is notable for its unusual stamens. The connective is elongate, filament-like, and jointed to the proper filament, with a pollen sac at each end, or with only the distal pollen sac functional.

222a (from 219b) Functional (anther-bearing) stamens as many as and alternate with the corolla lobes, usually four or five; corolla usually regular, seldom somewhat irregular 223

222b Functional stamens generally fewer than the corolla lobes, mostly two or four; corolla usually evidently irregular (except Oleaceae) 236

223a (from 222a) Stamens free from the corolla 224

223b Stamens borne on the corolla tube (or throat) 225

224a (from 223a) Flowers with a petaloid corona as well as corolla; mostly vines, climbing by tendrils; styles three and distinct, or one and deeply 3-cleft .... Passion-Flower Family, **PASSIFLORACEAE**

See fig. 150. Some species of Passiflora have the petals united at the base and might be sought here in the key.

224b Flowers without a corona; herbs or shrubs, not climbing; style one, entire or shortly lobed at the tip .... Heath Family, **ERICACEAE**

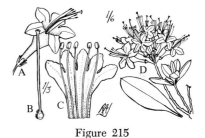

Figure 215

Figure 215 *Rhododendron occidentale* (T. & G.) A. Gray, California Azalea, is one of the several species of Ericaceae that have only five stamens. A, flower; B, pistil; C, corolla, laid open to show stamens; D, habit. Most members of the family have eight or ten stamens (twice as many as the corolla lobes) and would be sought elsewhere in the key. See fig. 204.

225a (from 223b) Ovary trilocular and stigmas or stigma lobes three 226

225b Ovary unilocular or bilocular and stigmas or stigma lobes two (or the stigma simple) 227

226a (from 225a) Corolla lobes imbricate in bud; evergreen perennial herbs or half-shrubs with simple, alternate leaves; flowers sometimes with five staminodes in addition to the five stamens .... **Diapensia Family,** **DIAPENSIACEAE**

Figure 216

Figure 216 *Diapensia lapponica* L., Diapensia, is a circumpolar species that reaches as far south as the higher mountains of New England and New York.

Galax, Pyxidanthera (Pyxie, Flowering Moss), and *Shortia* are some other American members of this small family, which has only about eighteen species.

226b Corolla lobes convolute in bud (each one with one edge inside and one outside); annual or perennial herbs or low shrubs, mostly not evergreen; leaves variously simple or compound, alternate or opposite; flowers without staminodes ...... Phlox Family, **POLEMONIACEAE**

Figure 217

Figure 217 A, *Phlox divaricata* L., Wild Blue Phlox; B, *Polemonium reptans* L., Jacob's Ladder; C, *Gilia congesta* Hook., Round-Head Gilia.

There are about 300 species of Polemoniaceae, nearly all of them in the New World. They are especially abundant in western United States. Many of them, notably species of *Gilia* and *Polemonium*, smell something like a skunk. Most members of the family can be quickly recognized by the 3-lobed or 3-parted style and sympetalous, regular flowers with a superior ovary and stamens alternate with the corolla lobes. Aside from the Polemoniaceae, only a few species of the Diapensiaceae meet this set of criteria.

227a (From 225b) Leaves nearly always opposite or whorled (alternate in *Buddleja alternifolia* 228

227b Leaves alternate (sometimes all basal) 232

228a (from 227a) Plants with milky juice 229

228b Plants without milky juice 230

229a (from 228a) Carpels united only by the thickened style-head, otherwise distinct; flowers usually with a well developed, petaloid corona in addition to the corolla; pollen grains coherent in pollinia; pollinia of the adjoining halves of adjacent anthers yoked by a translator, by means of which they may be extracted from the anthers in pairs .... Milkweed Family, ASCLEPIADACEAE

Figure 218

Figure 218   A, *Asclepias tuberosa* L., Butterfly-Weed; B, *Asclepias incarnata* L., Swamp Milkweed; C, *Asclepias viridiflora* Raf., Green Milkweed; D, *Gonolobus gonocarpus* (Walter) Perry, Angle-Pod.

There are about 2,000 species of Asclepiadaceae, most of them tropical. Asclepias is famous as the special host for caterpillars of the Monarch Butterfly, which are adapted to withstand the potent poison in the milky juice. *Ceropegia, Hoya,* and *Stapelia* are occasionally grown as house plants. There is no absolutely sharp dividing line between the Asclepiadaceae and Apocynaceae, but the species in the United States are easy enough to place in one family or the other. The highly complex flowers of the Asclepiadaceae are readily recognizable.

229b Carpels united by part or all of the style below the common style head, or more completely united; flowers without a corona; pollen grains not forming pollinia ....
...... Dogbane Family, APOCYNACEAE

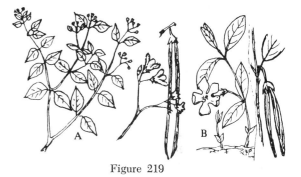

Figure 219

Figure 219   A, *Apocynum androsaemifolium* L., Spreading Dogbane; B, *Vinca minor* L., Periwinkle.

There are about 2,000 species of Apocynaceae, most of them tropical. *Allamanda* and *Nerium* (Oleander) are well known ornamental plants; both are poisonous. A single leaf of Oleander, if eaten, can be fatal even to an adult, and severe poisoning has resulted from the use of Oleander branches as skewers or roasting sticks in outdoor cookery. *Rauvolfia* and *Strophanthus* yield medically useful drugs. The Apocynaceae and Asclepiadaceae are unusual in that the carpels are usually separate toward the base, but united at the summit.

230a (from 228b) Ovary with two locules and axile placentas; leaves often with interpetiolar stipules 231

230b Ovary unilocular, with intruded parietal placentas; stipules wanting ....
.... Gentian Family, GENTIANACEAE

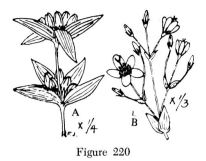

Figure 220

Figure 220   A, *Gentiana andrewsii* Griseb., Closed Gentian; B, *Sabatia angularis* (L.) Pursh, Marsh Pink.

There are about a thousand species of Gentianaceae, many of them in the North Temperate Zone. *Gentiana,* with some 400 species, is the largest genus. *Gentiana crinita* Froel., the Fringed Gentian, is another familiar and beautiful species.

231a (from 230a) Stamens and corolla lobes each five ....
.... **Logania Family, LOGANIACEAE**

Figure 221

Figure 221   *Gelsemium sempervirens* (L.) Ait.f., Yellow Jessamine, is the state flower of South Carolina. All parts of the plant contain poisonous alkaloids.

There are about 500 species of Loganiaceae, most of them tropical, only a few in the United States. The largest genus is *Strychnos,* the source of strychnine and some of the alkaloids of curare.

231b Stamens and corolla lobes each four .... **Butterfly Bush Family, BUDDLEJACEAE**

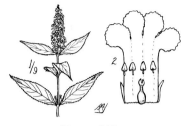

Figure 222

Figure 222   *Buddleja davidii* Franchet, Butterfly Bush, is a native of China that is often cultivated in the United States.

There are about 150 species of Buddlejaceae, most of them tropical or subtropical, about a hundred in the genus *Buddleja.*

232a (from 227b) Aquatic or semi-aquatic herbs; ovary unilocular, with parietal placentas .... **Buckbean Family, MENYANTHACEAE**

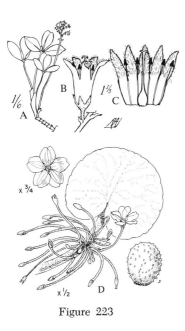

Figure 223

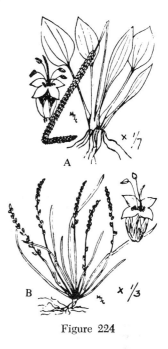

Figure 224

Figure 223 A-C, *Menyanthes trifoliata* L., Buckbean, is a circumboreal species of bogs and marshes. A, habit; B, flower; C, flower, laid open. D, *Nymphoides aquaticum* (Walter) Kuntze, Floating Heart, with floating leaves, grows in quiet water on the Atlantic Coastal Plain.

There are only about thirty species of Menyanthaceae, twenty of them in the genus *Nymphoides*. *Nymphoides* is named for its resemblance to *Nymphaea*, the Water Lily, in asaspect and habitat, but the flowers are very different.

232b Terrestrial herbs or shrubs or vines; ovary and placentas various 233

233a (from 232b) Corolla scarious, persistent, inconspicuous, usually 4-lobed; leaves usually all basal and with several prominent, nearly parallel veins .... Plantain Family, PLANTAGINACEAE

Figure 224 A, *Plantago major* L., Common Plantain, is a familiar weed in lawns and waste places, introduced from Europe; B, *Plantago elongata* Pursh, Slender Plantain, is native in western United States.

There are about 250 species of Plantaginaceae, nearly all in the genus *Plantago*. *Plantago lanceolata* L., the English Plaintain, is another common introduced weed in much of the United States. Children use its leaves to determine the number of lies recently told. If the leaf is pulled apart near the middle, crosswise to the main veins, some of the veins will stand out beyond the broken leaf tissue, indicating the number and magnture of lies.

233b Corolla otherwise; leaves rarely all basal, and then not appearing to be parallel-veined 234

234a (from 233b) Ovary with two locules and axile placentas 235

234b Ovary with a single locule and two parietal (sometimes intruded) placentas; style usually more or less deeply bifid .... Waterleaf Family, **HYDROPHYLLACEAE**

Figure 225

Figure 225   A, *Ellisia nyctelea* L., Nyctelea, is a common, somewhat weedy American species; B, *Hydrophyllum virginianum* L., Virginia Waterleaf, is widespread in woods in the eastern United States; C, *Nama ovatum* (Nutt.) Britton, grows in southwestern United States.

There are about 300 species of Hydrophyllaceae, many of them in western United States. About half the species belong to the genus *Phacelia*. Members of the Hydrophyllaceae are readily distinguished from the closely related family Polemoniaceae by the bifid rather than trifid style.

235a (from 234a) Ovules and seeds mostly two in each locule; style simple or more often more or less deeply cleft, or the styles two and distinct; plants mostly twiners or climbers .... Morning Glory Family, **CONVOLVULACEAE**

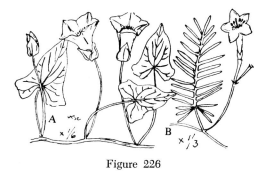

Figure 226

Figure 226   A, *Convolvulus sepium* L., Hedge Bindweed, is a native American twiner; B, *Ipomoea quamoclit* L., Cypress Vine, is an ornamental climber from tropical America. The Sweet Potato (*Ipomoea batatas* (L.) Lam.) and Morning Glory (spp. of *Ipomoea*) belong to this family.

There are about 1,500 species of Convolvulaceae. The largest genera are *Ipomoea* (400 species) and *Convolvulus* (250 species). Some of the Mexican species of *Ipomoea* are small, stout trees instead of twining vines, but the showy flowers have the characteristic structure and appearance of Morning Glories.

235b Ovules and seeds mostly more or less numerous in each locule; style simple, the stigma entire or only shortly bilobed; most species erect, but a few twining or climbing. (Note that *Verbascum*, of the Scrophulariaceae, with five stamens and only slightly irregular corollas, might be sought here; most species of Solanaceae have a regular corolla) .... Potato Family, **SOLANACEAE**

Figure 227

Figure 227 A, *Physalis alkekengi* L., Chinese Lantern Plant; B, *Solanum nigrum* L., Black Nightshade; C, *Datura stramonium* L., Jimson-Weed.

There are about 2,300 species of Solanaceae, 1,500 of them in the immense genus *Solanum*. *Solanum tuberosum* L. (Potato), *Lycopersicon esculentum* Miller (Tomato), and *Nicotiana tabacum* L. (Tobacco) are familiar members of this family. Many of the Solanaceae contain poisonous alkaloids, and even potatoes become poisonous when they develop green color as a result of exposure to light. The name Jimson-weed is said to be a modification of Jamestown Weed, referring to purported early colonial use of the highly poisonous seeds of *Datura stramonium* L. as a hallucinogen. Its leaves and even the nectar of its flowers are also very poisonous.

236a (from 222b) Corolla 4-lobed and regular or nearly so; stamens two; woody plants with opposite leaves .... Olive Family, OLEACEAE

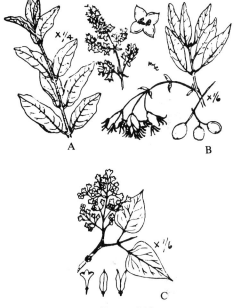

Figure 228

Figure 228 A, *Ligustrum vulgare* L., Privet, used in hedges; B, *Chionanthus virginica* L., Fringe Tree; C, *Syringa vulgaris* L., Common Lilac, the state flower of New Hampshire.

There are about 500 species of Oleaceae. The largest genus is *Jasminum*, with about 200 species. *Forsythia* is another familiar genus. The cultivated olive (*Olea*) also belongs here. In addition to the features mentioned in key lead 236a, only the superior, unlobed ovary is necessary to specify this family. See also fig. 56.

236b Corolla mostly 5-lobed and more or less strongly irregular; stamens two or four (rarely five); herbs or woody plants with opposite or alternate or whorled leaves (some cultivated plants, such as *Sinningia,* in the Gesneriaceae, may have larger numbers of stamens and corolla lobes) 237

237a (from 236b) **Fruit explosively dehiscent, the seeds with an enlarged and specialized funiculus that is typically developed into a jaculator** .... **Acanthus Family, ACANTHACEAE**

Figure 229

Figure 229   A, *Ruellia caroliniensis* (Walter) Steudel, Hairy Ruellia; B-F, *Aphelandra squarrosa* Nees, a Brazilian species that is sometimes grown as a house plant in the United States. B, habit; C, inflorescence; D, anther; E, long-section of flower; F, two views of ovary.

There are about 2,500 species of Acanthaceae, the vast majority of them tropical. Many have beautiful flowers, but only a few are suitable for cultivation in temperate climates. *Beloperone guttata* Brandegee, Mexican Shrimp Plant, is sometimes cultivated indoors or out in the United States. Aside from the specialized dehiscence-mechanism, the Acanthaceae are much like the Scrophulariaceae.

237b **Fruit indehiscent or dehiscent, but not explosively so; funiculus of ordinary type, not developed into a jaculator** .. 238

238a (from 237b) **Trees, shrubs, or woody vines; seeds commonly winged; leaves simple or often compound, the terminal leaflet(s) sometimes modified into tendrils** .... **Trumpet-Creeper Family, BIGNONIACEAE**

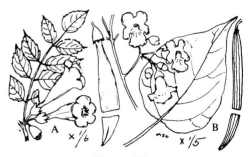

Figure 230

Figure 230 A, *Campsis radicans* (L.) Seem., Trumpet-Creeper, is a vine of southeastern United States with large, vermilion flowers; B, *Catalpa speciosa* Warder, Northern Catalpa, is a tree with large, heart-shaped leaves and long, slender pods with winged seeds.

There are about 700 species of Bignoniaceae, most of them tropical. *Spathodea* (Flame Tree, African Tulip Tree) and *Jacaranda* are familiar tropical street trees. *Paulownia* (Empress Tree) is hardy in much of the United States; it has leaves like those of Catalpa, but the pods are much shorter.

238b Herbs or seldom shrubs; seeds winged or much more often wingless; leaves simple or sometimes compound, without tendrils 239

239a (from 238b) Herbage slimy because of the specialized, mucilaginous hairs; fruits very often with hooks or horns or prickles, or sometimes winged .... **Sesame Family, PEDALIACEAE**

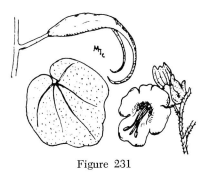

Figure 231

Figure 231 *Proboscidea louisiana* (Miller) Woot. & Standl., Unicorn Plant is one of the few species of this small (70 species) family native to the United States.

The largest genus in the family is *Sesamum*, with about twenty species. Sesame Seed comes from *Sesamum indicum* L.

239b Herbage smooth or hairy or sometimes glandular, but without specialized, mucilaginous hairs; fruit without hooks, horns, prickles or wings 240

240a (from 239b) Placentation parietal, but the placentas more or less deeply intruded; ovary superior or often partly inferior; plants sometimes of unusual vegetative structure, the shoot not clearly divisible into stems and leaves, but also sometimes of very ordinary structure .... **Gesneriad Family, GESNERIACEAE**

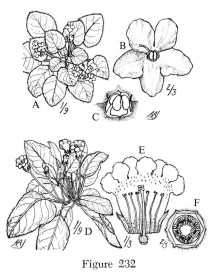

Figure 232

Figure 232 A-C, *Saintpaulia ionantha* H. Wendl., African Violet. A, habit; B, frontal view of flower; C, detail of corolla-throat, showing two stamens and two tiny staminodes. D-F, *Sinningia speciosa* (Lodd.) Hiern., Gloxinia, a native of Brazil. D, habit; E, flower, laid open; F, diagrammatic cross-section of ovary, showing parietal placentation. Wild plants of *Sinningia* have flowers with four stamens and an irregular 5-lobed corolla, like other gesneriads. Cultivated forms of *S. speciosa*, on the other hand, have been selected for a more nearly regular corolla with more numerous lobes. The flower shown here, obtained from a commercial nursery, has six stamens and seven corolla lobes.

There are about 2,000 species of Gesneriaceae, nearly all of them tropical. A few have become familiar house plants, and many others are grown by devoted amateur horticulturists.

240b Placentation axile; ovary superior; plants of ordinary vegetative structure **241**

241a (from 240b) Ovary with a single locule and ovule; fruit an achene, enclosed within the calyx, of which the three upper lobes become firm and hooked at the tip ....
...... **Verbena Family, VERBENACEAE**

Figure 233

Figure 233 *Phryma leptostachya* L., Lopseed, is an unusual member of the Verbenaceae, which has often been considered to form a separate family Phrymaceae. In spite of its distinctive pistil, it is much like *Verbena* in aspect. See fig. 213.

241b Ovary with two locules and several or usually numerous ovules; fruit usually a capsule, rarely a berry; calyx lobes not hooked .... **Figwort Family, SCROPHULARIACEAE**

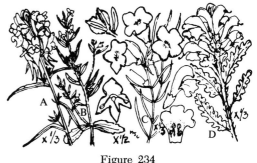

Figure 234

Figure 234 A, *Linaria vulgaris* L., Butter and Eggs, is a European species that has become well established as a casual weed in the United States; B, *Veronica peregrina* L., Purslane Speedwell, in contrast, is native to the New World and introduced into Europe. C, *Gerardia purpurea* L., Purple Gerardia. D, *Pedicularis canadensis* L., Canada Lousewort. These latter two species are native to eastern United States.

There are about 4,000 species of Scrophulariaceae, many of them with beautiful flowers. Species of *Castilleja* (Indian Paintbrush), *Penstemon* (Beard-tongue), *Mimulus* (Monkey-flower), and *Pedicularis* (Lousewort), in particular, provide many showy wild flowers in western United States. *Castilleja linariaefolia* Benth. is the state flower of Wyoming. The large high-Andean genus *Calceolaria* provides several spectacular potted plants, which require cooler, moister air than is found in most American homes. Several genera of Scrophulariacae, including *Castilleja*, are partial parasites, which attach to the roots of other plants but also have green leaves and make at least some of their own food. Only a few are non-green and wholly parasitic, as in the related family Orobanchaceae. In the United States, most species of plants with ordinary green leaves, a sympetalous, irregular corolla, two or four stamens, and a superior ovary that is not lobed, belong to the Scrophulariaceae.

**242a (from 218b) Tendril-bearing vines; leaves alternate; anthers generally fewer than the corolla lobes, most commonly three, two of them bilocular, the other unilocular ....
.... Squash Family, CUCURBITACEAE**

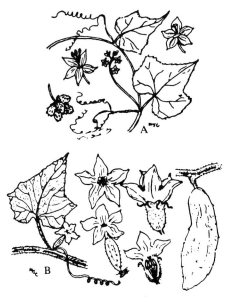

Figure 235

Figure 235 A, *Sicyos angulatus* L., Star Cucumber, is native to eastern and central United States; B, *Cucumis sativus* L., Cucumber, is a cultigen from southern Asia.

There are about 700 species of Cucurbitaceae, most of them tropical. Pumpkins, squashes, gourds, and watermelons belong to this family. Most tendril-bearing vines with a sympetalous corolla and an inferior ovary belong to the Cucurbitaceae.

242b Plants without tendrils, although sometimes with twining stems; leaves alternate or opposite; anthers fewer than or as many as the corolla lobes, generally all bilocular 243

**243a (from 242b) Flowers grouped into heads, each head subtended by an involucre of bracts (reduced, modified leaves) and often resembling an individual flower; calyx modified to form a pappus of scales or hairs or hooks or bristles or a chaffy ring, or wanting; flowers with a specialized pollen-presentation mechanism, the anthers elongate, connate by their margins to form a tube into which the pollen is released, the pollen pushed out of the tube by the growth of the style; ovary with a single locule and a single basal ovule, but the style generally with 2 branches; plants herbaceous or less often woody, with alternate or opposite leaves ....
.... Aster Family, ASTERACEAE**

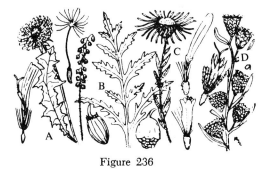

Figure 236

Figure 236   A, *Taraxacum officinale* Weber, Common Dandelion, comes originally from Europe; B, *Ambrosia artemisiifolia* L., Common Ragweed, is a native American weed; C, *Erigeron caespitosus* Nutt., Tufted Daisy, is one of many dwarf species of *Erigeron* in the western cordillera; D, *Liatris cylindracea* Michx., Blazing Star, is a common middle-western species; other species of *Liatris* occur throughout the area of the United States that lies east of the Rocky Mountains.

The characteristic flower-heads and specialized pollen-presentation mechanism set the Asteraceae apart from nearly all other angiosperms. The family has nearly 20,000 species and is found throughout the world. Many are familiar ornamentals, such as Aster, Chrysanthemum, Dahlia, Daisy, Marigold, Sunflower, and Zinnia. The family provides the state flower for several states: Goldenrod (*Solidago*), for Nebraska and Kentucky; Black-eyed Susan (*Rudbeckia hirta* L.) for Maryland; Sagebrush (*Artemisia tridentata* Nutt.) for Nevada; and Sunflower (*Helianthus annuus* L.) for Kansas. Lettuce (*Lactuca sativa* L.), Sunflower, and Artichoke (*Cynara scolymus* L.) are among the few food plants in the family.

243b **Flowers otherwise, sometimes grouped into heads, but then without the specialized pollen-presentation mechanism**
of the Asteraceae; other features various 244

244a (from 243b) Leaves alternate; filaments attached to the base of the corolla tube, or free from the corolla; flowers with a specialized pollen-presentation mechanism resembling that of the Asteraceae 245

244b Leaves opposite or whorled; filaments attached well up in the corolla tube; flowers usually without the previously described pollen-presentation mechanism 246

245a (from 244a) Style with a more or less cupulate structure just beneath the small stigma, but without collecting hairs .... Goodenia Family, **GOODENIACEAE**

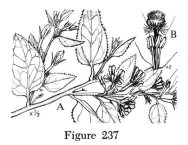

Figure 237

Figure 237   *Goodenia ovata* Smith, Hop Goodenia, an Australian shrub with yellow flowers. A, habit; B, detail of part of flower, with stamens and distally enlarged style.

There are about 300 species of Goodeniaceae, most of them Australian, none native to the United States, none familiar in cultivation. A few species of *Scaevola* are mangroves.

245b Style provided with a ring of collecting hairs below the 2-3 (5) initially appressed stigmas, but without a subtending cupulate structure .... Harebell Family, CAMPANULACEAE

Figure 238

Figure 238 A, *Campanula rotundifolia* L., Bluebell of Scotland, is a circumboreal species; B, *Triodanis perfoliata* (L.) Nieuwl., Venus' Looking-Glass, is a widespread, rather weedy American species; C, *Lobelia cardinalis* L., Cardinal Flower, is native in wet places throughout eastern United States.

There are about 2,000 species of Campanulaceae. The largest genera are *Campanula*, with regular, bell-shaped flowers, and *Lobelia*, with highly irregular flowers. Among plants that grow wild in the United States, the Campanulaceae are easily recognized by the combination of alternate leaves, sympetalous corolla, five stamens, inferior ovary, and relatively open inflorescence (unlike the compact heads of the Asteraceae). The genera with irregular flowers are often treated as a separate family Lobeliaceae.

246a (from 244b) Leaves with interpetiolar stipules, or the leaves whorled; corolla usually regular; stamens as many as the corolla lobes (typically four or five) .... Madder Family, RUBIACEAE

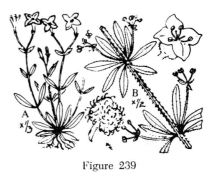

Figure 239

Figure 239 A, *Houstonia caerulea* L., Bluets, is a common spring flower in eastern United States; B, *Galium aparine* L., Bedstraw, is a weed introduced from Europe.

There are about 6,500 species of Rubiaceae, most of them tropical woody plants. Most of the species in the temperate zone are herbaceous. *Galium*, with whorled leaves, is the largest genus in the family in temperate regions. *Cinchona*, the source of quinine, and *Coffea*, the source of coffee, are economically important genera.

246b Leaves mostly without stipules, never whorled; corolla sometimes regular, but more often irregular; stamens as many as or often fewer than the corolla lobes 247

247a (from 246b) Plants mostly woody, seldom herbaceous; stamens mostly as

many as the corolla lobes, seldom fewer
.... Honeysuckle Family,
CAPRIFOLIACEAE

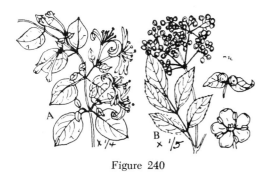

Figure 240

Figure 240 A, *Lonicera japonica* Thunb., Japanese Honeysuckle, has become a common
weed in much of the United States; B, *Sam-
bucus canadensis* L., American Elderberry, is
common in eastern United States; other species of Elderberry occur farther west.

There are about 400 species of Caprifoliaceae,
most of them in North Temperate or Boreal
regions. The largest genera are *Lonicera* and
Viburnum, with about 150 species each. *Symphoricarpos* (Snowberry, Coralberry) and
Abelia are also familiar in cultivation. The
Caprifoliaceae are closely related to the Rubiaceae on one hand, and the Valerianaceae and
Dipsacaceae on the other.

247b Plants mostly herbaceous, rarely shrub-
     by; stamens usually fewer than the cor-
     olla lobes 248

248a (from 247b) Ovules and seeds numer-
     ous .... Gesneriad Family,
     GESNERIACEAE

See fig. 232. Some genera of Gesneriaceae have
a more or less inferior ovary and would be
sought here.

248b Ovules and seeds not more than one
     per locule 249

249a (from 248b) Flowers individually en-
     closed or subtended by a more or less
     cupulate epicalyx or involucel, mostly
     borne in compact, basically cymose
     heads, each of which is subtended by
     an involucre of bracts; ovary strictly
     unilocular ....
     .... Teasel Family, DIPSACACEAE

Figure 241

Figure 241 *Dipsacus sylvestris* Huds., Common Teasel, is a weed introduced from Europe. Its cultivated close relative, *Dipsacus
fullonum* L., is the Fuller's Teasel, traditionally used to raise the nap on cloth.

There are nearly 300 species of Dipsacaceae,
native to Eurasia and Africa. Several species
of *Scabiosa,* the Pincushion Flower, are garden ornamentals. The flower-heads of Dipsacaceae are something like those of the Asteraceae (though not in detail), but the flowers
have separate anthers, not joined into a tube.

249b Flowers without an epicalyx or involu-
     cel, borne in various sorts of inflores-
     cences, but not in involucrate heads;
     ovary more or less evidently trilocular,
     with one fertile and two sterile locules
     .... Valerian Family, VALERIANACEAE

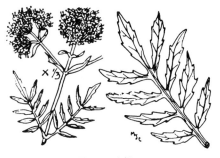

Figure 242

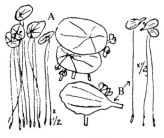

Figure 243

Figure 242 *Valeriana officinalis* L., called Garden Heliotrope, is somewhat misnamed, because it has nothing to do with *Heliotropium*, in the Boraginaceae. Several species of *Valeriana* are native to the United States, but *V. officinalis* comes from Eurasia.

There are about 300 species of Valerianaceae, many of them in North Temperate regions. Only a few species are occasionally grown as ornamentals.

Class Liliopsida, The Monocotyledons

250a (from 11b) Plants aquatic, either free-floating or rooted to the bottom; if rooted to the bottom, then variously submerged or emergent or with floating leaves 251

250b Plants mostly terrestrial (or epiphytic), sometimes emergent from shallow water, but then differing in other respects from each family of the preceding group 267

251a (from 250a) Plants thalloid, free-floating, small, the shoot not differentiated into stem and leaves, the root(s) short or wanting; flowers much reduced, rarely produced .... Duckweed Family, LEMNACEAE

Figure 243 A, *Spirodela polyrhiza* (L.) Schleiden, Greater Duckweed; B, *Lemna trisulca* L., Ivy-leaved Duckweed.

The Lemnaceae are a cosmopolitan group of about thirty species. They form green carpets of myriad individuals on quiet inland waters. Botanists consider that the Duckweeds are a reduced, aquatic offshoot of the Araceae. *Pistia*, a free-floating tropical aroid, is often cited as a connecting form.

251b Plants with evident leaves attached to short or elongate stem; roots and flowers various 252

252a (from 251b) Carpels united to form a compound ovary; perianth well developed, with definite petals; ovules several or many 253

252b Carpels separate, or seemingly or actually only one, or sometimes united to form a compound ovary, but then the flowers without evident petals; ovules 1-many 255

253a (from 252a) Ovary inferior .... Frog's bit Family, HYDROCHARITACEAE

Figure 244

Figure 244 A, *Elodea canadensis* Richard, Water-weed; B, *Vallisneria americana* Michx., Tape-grass. Both species are native to quiet waters in the United States; both are used as aerators in aquaria, but the South American species *Elodea densa* Planchon is more commonly used for this purpose.

There are about a hundred species of Hydrocharitaceae, the majority of them tropical. *Elodea* is famous for its unusual mechanism of pollination. The flowers are unisexual and borne at the surface of the water, the pistillate ones actually with a submerged, sessile ovary and a slender, elongate, pedicel-like hypanthium extending to the water surface. The staminate flowers of some species break off and float about. Pollination depends on the casual contact of the anthers of these detached staminate flowers with the enlarged stigmas of the pistillate flowers. In other species of *Elodea* the pollen is released onto the water surface and floats about.

253b Ovary superior **254**

254a (from 253b) Sepals and petals separate, free, the sepals green, not petaloid .... Mayaca Family, **MAYACACEAE**

Figure 245

Figure 245. *Mayaca aubletii* Michx., has small white flowers. It grows in still waters in southeastern United States.

There are only four species of Mayacaceae, all of warm regions.

254b Sepals and petals both petaloid, united below to form a perianth tube .... **Pickerel-weed Family,** **PONTEDERIACEAE**

Figure 246

Figure 246. A, *Pontederia cordata* L., Pick-erel-weed, of eastern United States, is a coarse, colonial plant, rooted to the bottom, with a cylinder of blue flowers above the water; B, *Eichhornia crassipes* (Mart.) Solms., Water Hyacinth, is tropical and subtropical. It has inflated petioles and is free-floating; the flowers are lavender.

There are about thirty species of Pontederia-ceae, most of them tropical or subtropical.

255a (from 252b) Perianth differentiated into evident sepals and petals, or the sepals colored like the petals **256**

255b Perianth small and inconspicuous, or wanting, not evidently differentiated into sepals and petals **258**

256a (from 255a) Ovules one or two, sel-dom more, on basal or marginal pla-centas .... Water Plantain Family, **ALISMATACEAE**

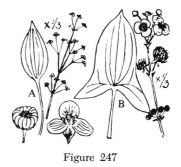

Figure 247

Figure 247 A, *Alisma plantago-aquatica* L., Water Plantain, is a common cosmopolitan species; B, *Sagittaria latifolia* Willd., Broad-leaved Arrowhead, is widespread in North America.

The family Alismataceae consists of some sev-enty species, growing rooted in marshes and quiet water. They have emergent, generally white flowers and basal leaves that usually have a long petiole and a floating or emergent blade; sometimes the leaves are wholly sub-merged and ribbon-shaped.

256b Ovules several or many, scattered over the inner surface of the carpel **257**

257a (from 256b) Leaves linear, not differ-entiated into blade and petiole; petals peristent; sepals somewhat petaloid .... Flowering-rush Family, **BUTOMACEAE**

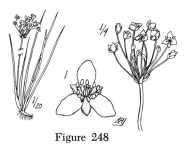

Figure 248

Figure 248 *Butomus umbellatus* L., Flowering Rush.

The family consists of a single Eurasian species, introduced in eastern United States along shores and river banks.

257b Leaves with a blade and petiole; petals quickly deciduous; sepals green .... Water-poppy Family, LIMNOCHARITACEAE

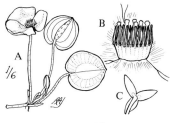

Figure 249

Figure 249 *Hydrocleys nymphoides* (Willd.) Buchenau, Water Poppy, with floating leaves and yellow flowers with purple stamens, is native to tropical America and is often cultivated in aquaria. A, habit; B, detail of part of flower, showing bases of petals, stamentube, and styles; C, calyx and pedicel, from beneath.

There are only about a dozen species of Limnocharitaceae, all tropical or subtropical.

258a (from 255b) True aquatics with submersed or floating leaves, only the inflorescence sometimes (Potamogetonaceae) emergent 259

258b Emergent marsh-plants (sometimes only the inflorescence emergent in Sparganiaceae); occasional members of certain other families (notably Araceae, Cyperaceae, Eriocaulaceae, Juncaceae, and Xyridaceae) might be sought here, but do not have the features given for the individual families of this group 264

259a (from 258a) Plants truly marine, growing rooted to the bottom in shallow water near the shore; flowers unisexual 260

259b Plants growing in fresh or brackish water, not truly marine; flowers variously perfect or unisexual 261

260a (from 259a) Flowers in small axillary cymes, or solitary in the axils; anthers two, paired on a common stalk and united back to back, each anther with two pollen sacs .... Manatee Grass Family, CYMODOCEACEAE

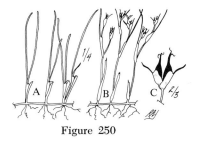

Figure 250

Figure 250 *Syringodium filiforme* Kuetz., Manatee Grass, grows in coastal waters in southern Florida and the West Indies. A, non-flowering plant; B, pistillate flowering plant; C, pair of pistillate flowers, with subtending bracts.

There are only about fifteen species of Cymodoceaceae, all native to tropical and subtropical sea coasts. Growing wholly submerged, they are inconspicuous and seldom noticed.

260b Flowers in spikes or clusters of spikes on axillary and/or terminal penduncles; stamen solitary, sessile, with a single pollen sac .... Eel Grass Family, ZOSTERACEAE

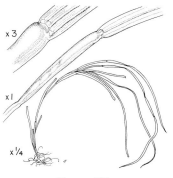

Figure 251

Figure 251 *Zostera marina* L., Eel Grass, is widespread along the sea shore from subarctic to subtropical regions. Although it usually grows in shallow water, sometimes it is found in water as much as 50 m. deep. Abundant in some places, it supports a great variety of marine animal life and is a staple winter food for some kinds of ducks and geese.

The family Zosteraceae has only about fifteen species, ten of them in the genus *Zostera*.

261a (from 259b) Pistil solitary, composed of a single carpel; leaves opposite or whorled .... Water-Nymph Family, NAJADACEAE

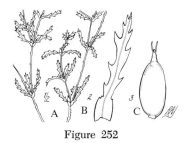

Figure 252

Figure 252 *Najas marina* L., Water Nymph, is nearly cosmopolitan. A, habit; B, single leaf; C, pistillate flower.

The family has about thirty-five species, which grow in quiet fresh or brackish water throughout most of the world. *Najas* is the only genus.

261b Pistils several, each composed of a single carpel; leaves variously alternate, opposite, or whorled 262

262a (from 261b) Flowers in axillary cymes, or solitary in the axils, unisexual .... Horned Pondweed Family, ZANNICHELLIACEAE

Figure 253

Figure 253 *Zannichellia palustris* L., Horned Pondweed, grows in fresh or brackish water throughout most of North America, and also in the Old World; it is the only New World species of its family.

There are only about seven species of Zannichelliaceae. None is of any great economic importance.

262b Flowers in spikes or racemes, perfect or less often unisexual **263**

263a (from 262b) Spikes terminal, submerged, at anthesis hidden by the leaf-sheath, in fruit with an elongate peduncle and stipitate achenes; tepals none; stamens two .... **Ditch-Grass Family, RUPPIACEAE**

Figure 254

Figure 254 *Ruppia maritima* L., Ditch Grass, is nearly cosmopolitan in coastal and brackish inland waters. It is the only species of Ruppiaceae.

263b Spikes axillary, emergent; tepals four; stamens four .... **Pondweed Family, POTAMOGETONACEAE**

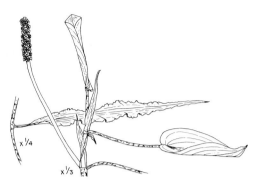

Figure 255

Figure 255 *Potamogeton pulcher* Tuckerman, Coastal Plain Pondweed, grows in quiet water, chiefly on the Atlantic Coastal Plain.

The family Potamogetonaceae consists of about ninety species, cosmopolitan in fresh water, often with relatively broad floating leaves and narrower submersed leaves.

264a (from 258b) Flowers perfect, or the basal ones rarely pistillate; carpels (1) three or six **265**

264b Flowers in unisexual spikes or heads; carpel apparently one or seldom two **266**

265a (from 264a) Seeds two in each of the three divergent pistils; inflorescence with a bract subtending each pedicel .... **Scheuchzeria Family, SCHEUCHZERIACEAE**

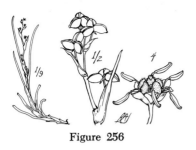

Figure 256

Figure 256 *Scheuchzeria palustris* L. is a circumboreal species of sphagnum bogs and lake margins. It is the only species of its family.

265b Seeds one in each of the (1) three or six erect, often more or less connate pistils; inflorescence usually bractless .... Arrowgrass Family, JUNCAGINACEAE

Figure 257

Figure 257 *Triglochin maritima* L., Arrowgrass, is a circumboreal species of both fresh and brackish marshes. It is poisonous to livestock.

The family has about eighteen species, widespread in temperate and cold regions.

266a (from 264b) Inflorescence of two dense, cylindrical, unisexual spikes on a continuous axis, the staminate spike above the pistillate one; achenes slenderly long-stipitate, with long hairs on the stipe, wind distributed ....
.... Cattail Family, TYPHACEAE

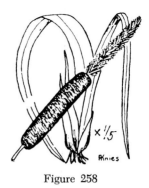

Figure 258

Figure 258 *Typha latifolia* L., Broad-leaved Cattail, is widespread in the Northern Hemisphere.

The family has only the genus *Typha*, with about fifteen species, abundant in marshes throughout most of the world. The genus is easily recognized by its characteristic habit and inflorescence.

266b Inflorescence of several globose, unisexual heads; achenes sessile or nearly so, not wind distributed .... Bur-reed Family, SPARGANIACEAE

Figure 259

Figure 259 *Sparganium eurycarpum* Engelm., Two-styled Bur Reed, is widespread in the United States in mud or shallow water.

There are about twenty species of *Sparganium*, the only genus of its family. Most of the species occur in temperate or boreal regions.

267a (from 250b) Leaves of palm type, large, with a sheath, petiole, and expanded, plicate (fan-folded) blade 268

267b Leaves otherwise, sometimes with a blade and petiole, but the blade not plicate 269

268a (from 267a) Flowers well developed (though individually rather small), usually with an evident, biseriate perianth of six members, not crowded into a spadix; plants often arborescent ....
.... Palm Family, ARECACEAE

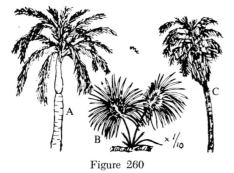

Figure 260

Figure 260 A, *Roystonea elata* (Bartram) F. Harper, the Florida Royal Palm, is native to Florida, where it is also often planted; B, *Sabal minor* (Jacq.) Pers., Dwarf Palmetto, grows in lowlands on the coastal plain from North Carolina to Florida and Texas; C, *Washingtonia robusta* H. Wendl, Thread Palm, is native to northwestern Mexico and is often grown in southern United States.

There are about 3,500 species of palms, all of warm regions, most of them strictly tropical. Most palms are trees with a slender, unbranched trunk and a terminal crown of large leaves, but some species are climbers, or have a very short (or no) trunk, the leaves arising directly from the ground.

268b Flowers reduced, with minute or no perianth, crowded into a spadix; plants usually herbaceous, never arborescent .... Cyclanthus Family, **CYCLANTHACEAE**

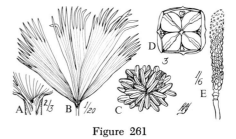

Figure 261

Figure 261 *Carludovica palmata* Ruiz & Pavon, the Panama Hat Plant, is the main source of fiber for Panama hats. A, base of leaf-blade; B, leaf-blade; C, staminate flower; D, pistillate flower, viewed from above; E, inflorescence, with staminate flowers above, pistillate below.

The Cyclanthaceae are a tropical American family of about 180 species that have palm-like, often bifid leaves, but aroid-like inflorescences.

269a (from 267b) Flowers reduced, the perianth wanting, or tiny and not differentiated into definite sepals and petals, in any case without evident petal-like parts 270

269b Flowers with a well developed, generally biseriate perianth, with one or both sets of tepals petal-like 276

270a (from 269a) Inflorescence a spadix (i.e., a spike with small, crowded flowers on a thickened, fleshy axis, the whole subtended by a specialized leaf called a spathe) 271

270b Inflorescence not a spadix 272

271a (from 270a) Woody plants with numerous firm, narrow, parallel-veined leaves that are usually spiny-margined; stamens 10-many; plants dioecious ....
.. Screw-pine Family, PANDANACEAE

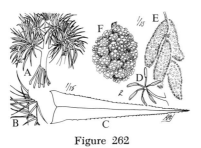

Figure 262

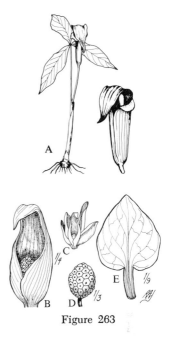

Figure 263

Figure 262   *Pandanus utilis* Bory, Common Screw Pine, is a native of Madagascar that is cultivated in tropical countries for ornament, food, and fiber. A, habit; B, detail showing spirals of leaves; C, leaf; D, staminate flower; E, staminate inflorescence; F, cone-like pistillate infloroscene.

There are several hundred species of Pandanaceae, all native to the Old World tropics, often along the sea shore. Most of them belong to the large genus *Pandanus*. The stem in Pandanaceae has a characteristic spiral growth-pattern, with three or four ranks of leaves spiraled about the stem.

271b Herbs or somewhat woody climbers; leaves otherwise, often with an expanded, net-veined blade; stamens 1-8; flowers perfect or unisexual, but the plants seldom dioecious ....
.... Aroid Family, ARACEAE

Figure 263   A, *Arisaema triphyllum* (L.) Schott, Jack-in-the-Pulpit, is native to forests of eastern United States; B-D, *Symplocarpus foetidus* (L.) Nutt., Skunk Cabbage, is one of the first flowers of spring in eastern United States. In its early-season growth it respires rapidly and converts much of the energy of respiration into heat, so that its temperature is substantially above that of its surroundings. B, spathe with included spadix; C, spathe and spadix with developing young leaves; D, spadix (inflorescence); E, mature leaf.

There are about 1,800 species of Araceae, most of them herbs of tropical and subtropical forests; relatively few species occur in temperate regions. *Philodendron* and *Dieffenbachia* (Dumb-cane) are in this family. Both of these genera are poisonous if eaten, and *Dieffenbachia* can cause the mouth and tongue to swell to the point of strangulation. *Anthurium, Caladium, Monstera,* and *Spathiphyllum* are some other genera of Araceae that are well known in cultivation.

272a (from 270b) Ovary with 2-many ovules; perianth generally present, though small and not petaloid; inflorescence various **273**

272b Ovary with a single ovule; flowers in characteristic spikes or spikelets, with or without an inconspicuous perianth .. **275**

273a (from 272a) Ovules solitary in each of the 2-3 locules of the ovary; flowers mostly unisexual **274**

273b Ovules 3-many in the single locule, or many in each of the three locules of the ovary; flowers mostly perfect, green or more often brown, in various sorts of inflorescences ....
.... **Rush Family, JUNCACEAE**

Figure 264 A, *Juncus tenuis* Willd., Path Rush, is widespread in North America, often along woodland paths; B-H, *Luzula campestris* L., Wood Rush, grows in temperate parts of both the Old and the New World. B, habit; C, portion of leaf sheath, surrounding the stem, and the base of the blade; D, inflorescence; E, opened capsule, showing three seeds; F, flower, subtended by two bractioles; G, flower, opened out to show perianth and stamens; H, pistil.

There are about 300 species of Juncaceae. They occur throughout much of the world, especially in temperate and boreal regions, but most of them are relatively inconspicuous, and they are not of much economic importance. They are much like the Liliaceae in aspect, except for the small, dry, chaffy perianth.

274a (from 273a) Flowers tightly clustered into dense, white to gray or lead-colored, involucrate heads terminating the stems; leaves all basal, without a well differentiated sheath .... Pipewort Family, **ERIOCAULACEAE**

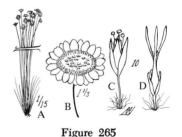

Figure 265

Figure 264

Figure 265 *Syngonanthus elegans* (Bong.) Ruhl., of tropical America, an "everlasting," with persistent white involucral bracts, is sometimes sold by florists. A, habit; B, flowerhead, with petal-like involucral bracts; C, staminate flower; D, pistillate flower.

There are about 600 species of Eriocaulaceae, most of them tropical, many in South America. The plants commonly have a cluster of narrow, grass-like basal leaves and a terminal head on a scape; the perianth is small, dry, and chaffy. Several species of *Eriocaulon* (Pipewort) native to shallow temporary pools in southeastern United States look much like *Syngonanthus* except that they do not have conspicuous involucral bracts.

274b Flowers borne in grass-like spikelets that are variously arranged; leaves mostly cauline, usually reduced to an open sheath with scarcely any blade .... Restio Family, RESTIONACEAE

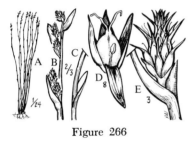

Figure 266

Figure 266 *Restio australis* R. Br., is a grasslike plant native to Australia. A, habit; B, inflorescence; C, portion of stem, showing reduced, sheathing leaf; D, functionally pistillate flower, with a bicarpellate pistil and a reduced stamen; D, pistillate spikelet.

The family Restionaceae has about 400 species, widely distributed in the Southern Hemisphere, best developed in Australia and South Africa. They play an ecological role similar to

that of the grasses, and botanists often refer to them as Southern Hemisphere grasses.

275a (from 272b) Flowers spirally or less often distichously arranged on the axis of the spike or spikelet, usually each flower seemingly or actually subtended by only a single bract, without an evident bract between the flower and the axis; leaf-sheath usually closed; stem usually solid, often triangular; flowers often with a perianth of evident bristles; carpels three or less often two .... Sedge Family, CYPERACEAE

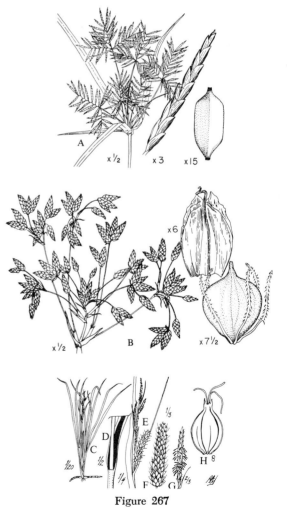

Figure 267

Figure 267 A, *Cyperus odoratus* L., Scented Cyperus, is widespread in tropical and warm-temperate regions; B, *Scirpus acutus* Muhl., Great Bulrush, grows in wet places and shallow water across most of the United States; C-H, *Carex rostrata* Stokes, Beaked Sedge, is a circumboreal species of wet places and shallow water. C, habit; D, portion of leaf and stem, showing closed sheath; E, inflorescence; F, pistillate spike; G, staminate spike; H, pistillate flower, with its enclosing bract, the perigynium. The perigynium is a special feature of *Carex*, not found in other Cyperaceae.

There are about 4,000 species of Cyperaceae, found throughout the world. The Cyperaceae are closely related to the Grass family, but are of relatively little economic importance.

275b Flowers distichously arranged on the axis of the spikelet (or only one per spikelet), each flower ordinarily subtended by a pair of bracts (lemma and palea), the palea inserted between the flower and the axis; leaf-sheath usually open; stem usually hollow, never triangular; flowers essentially without perianth; carpels generally two .... Grass Family, POACEAE

Figure 268

Figure 268 A-E, *Poa pratensis* L., Kentucky Bluegrass, is a circumboreal species much used in lawns and pastures. A, habit; B, portion of stem and leaf, with open sheath; C, portion of inflorescence; D, spikelet; E, single flower, with lemma (left), and palea (right), three stamens, and pistil. F, *Phragmites australis* (Cav.) Trin., Common Reed, grows in salt-marshes throughout the world. It is becoming frequent along highways, possibly because of the use of salt as a de-icer.

The grasses are a large, cosmopolitan, economically vital family of some 8,000 species, the source of cereals (Wheat, Maize, Rice, Oats, etc.), and the most important food for grazing animals. The family appears to have originated about at the beginning of the Tertiary geologic era, perhaps in association with the rise of grazing mammals. The intercalary meristem in the leaf at the base of the blade, just above the sheath, permits the leaf to continue to grow and function even after the tip has been nipped off. This same feature helps permit the grass in a lawn to withstand being mowed.

276a (from 269b) Stamens with anthers 3-6; flowers variously regular or irregular and with a superior or inferior ovary 277

276b Stamens with anthers only one or two; flowers always irregular and with an inferior ovary 292

277a (from 276a) Sepals not petaloid, only the petals showy 278

277b Sepals and petals both petaloid, though not always alike 284

278a (from 277a) Ovary superior; plants terrestrial, mesophytic to semi-aquatic 279

278b Ovary inferior to sometimes superior, in the latter case the plants epiphytic 281

279a (from 278a) Flowers individually large and usually showy, borne singly or in few-flowered, umbel-like inflorescences; petals usually more than 2 cm long, fairly firm and lasting several days; leaves in *Trillium* a single whorl beneath the solitary flower, sessile, relatively broad and more or less net-veined, in *Calochortus* mostly at or near the base, more slender, and parallel-veined .... Lily Family, LILIACEAE

Figure 269

Figure 269 *Trillium grandiflorum* (Michx.) Salisb., Large-flowered Wake-Robin, is a common spring flower in woods in eastern United States; other species occur in the West, as well as in the East. *Trillium* and *Calochortus* (Mariposa Lily) are our only genera of Liliaceae that will key here. The other genera have petal-like sepals and will key elsewhere. See fig. 283. *Calochortus nuttallii* Torrey and Gray, the Sego Lily, is the state flower of Utah.

279b Flowers smaller and more ephemeral, the delicate petals up to about 2 cm long 280

280a (from 279b) Leaf-sheath open, often not well differentiated from the narrow, grass-like blade; inflorescence a simple, racemose head terminating a long peduncle or scape, the leaves commonly mostly or all in a basal cluster; flowers most often yellow .... Yellow-eyed Grass Family, XYRIDACEAE

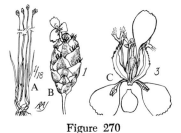

Figure 270

Figure 270. *Xyris caroliniana* Walter, Carolina Yellow-eyed Grass, is common in wet, low places in eastern United States. A. habit; B, inflorescence, with a single flower open; C, single flower, with lower sepal detached.

There are nearly 300 species of Xyridaceae, most of them in tropical and subtropical regions.

280b Leaves differentiated into a closed sheath and a well defined, commonly somewhat succulent blade; inflorescence cymose (or the flower solitary), axillary to a spathe or foliaceous bract; stem generally leafy; flowers mostly blue or pink or white .... Spiderwort Family, COMMELINACEAE

Figure 271

Figure 271 A, *Commelina communis* L., Asiatic Day-flower, is a common house plant, cultivated for its showy, often striped leaves; B, *Tradescantia ohiensis* Raf., Ohio Spiderwort, grows in moist woods and on prairies in eastern United States.

There are about 600 species of Commelinaceae, most of them tropical or subtropical.

281a (from 278b) Stamens six; flowers regular or somewhat irregular; xerophytes and epiphytes with narrow, often firm and spiny-margined leaves ....
.. Pineapple Family, BROMELIACEAE

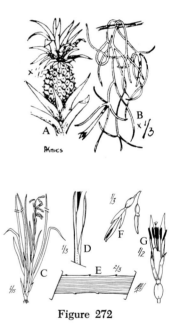

Figure 272

Figure 272 A, *Ananas comosus* (L.) Merrill, Pineapple, originated in South America, but is now widely cultivated in tropical countries; B, *Tillandsia usneoides* L., Spanish Moss, commonly drapes on branches of tree throughout tropical America, north to the southeastern United States; C-G, *Billbergia nutans* H. Wendl. Friendship Plant, from Brazil, is common in cultivation under glass. C, habit; D, portion of stem with leaf-base; E, portion of leaf; F, flower and developing bud; G, flower, with one petal partly removed.

There are about 1,800 species of Bromeliaceae, almost all of them native to tropical America. The majority are epiphytes. Many species are cultivated under glass by a devoted coterie of amateur horticulturists.

281b Stamens five (six only in *Ravenala*, of the Strelitziaceae); flowers distinctly irregular; mesophytes with unarmed, pinnately veined, often broad leaves **282**

282a (from 281b) Flowers unisexual; leaves and bracts spirally arranged; fruit fleshy, indehiscent ....
.................**Banana Family, MUSACEAE**

Figure 273

Figure 273 *Musa paradisiaca* L., Banana, is a hybrid cultigen, perhaps originally from India. It is now grown in moist, tropical parts of the New World as well.

There are about forty species of Musaceae, all native to tropical regions in the Old World. Manila hemp comes from the sheathing petioles of *Musa textilis* Née.

282b Flowers perfect; leaves and bracts distichously arranged (in two opposite rows); fruit dry, dehiscent or separating into its component carpels **283**

283a (from 282b) Ovules numerous in each locule; fruit capsular; seeds arillate; stigma trifid .... **Bird-of-Paradise Flower Family, STRELITZIACEAE**

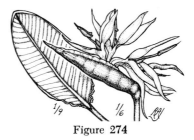

Figure 274

Figure 274 *Strelitzia reginae* Aiton, Bird-of-Paradise Flower, originally from South America, is cultivated indoors in the United States for its spectacular flowers.

There are only seven species of Strelitziaceae, all tropical. *Ravenala madagascariensis* Sonn., of Madagascar, is called Traveler's Tree because the cup-like leaf bases hold water that travelers may drink.

283b Ovules solitary in each locule; fruit schizocarpic (i.e., the carpels separating at maturity); seeds not arillate; stigma capitate .... **Heliconia Family, HELICONIACEAE**

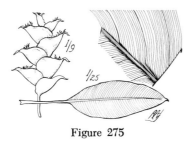

Figure 275

Figure 275 *Heliconia caribaea* Lam., Wild Plantain, is cultivated in tropical regions and under glass northward for the beauty of both its leaves and its flowers.

The family consists of the single genus *Heliconia,* with a hundred or more species, most of them native to tropical America.

284a (from 277b) Seeds very numerous and tiny, with undifferentiated embryo and little or no endosperm; ovary inferior; plants very often without chlorophyll .... **Burmannia Family, BURMANNIACEAE**

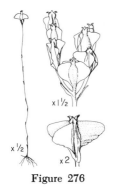

Figure 276

Figure 276 *Burmannia biflora* L., Blue Burmannia, grows in bogs on the coastal plain of southeastern United States.

The Burmanniaceae are a small, chiefly tropical, highly mycotrophic family of some 130 species, only a few in the United States.

284b **Seeds of ordinary number and structure, usually with a well differentiated embryo and well developed endosperm; ovary superior to inferior; plants with chlorophyll** 285

285a **(from 284b) Plants mostly climbing, very often dioecious; leaves with a definite petiole and an expanded (often large), more or less net-veined blade** 286

285b **Plants not climbing, usually with perfect flowers; leaves typically narrow and parallel-veined, sometimes broader and more net-veined, but only seldom with both a broad, net-veined blade and a distinct petiole** 288

286a **(from 285a) Flowers dimerous** .... **Stemona Family, STEMONACEAE**

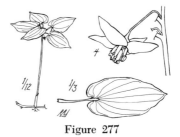

Figure 277

Figure 277 *Croomia pauciflora* (Nutt.) Torr., grows in Florida and Georgia.

The Stemonaceae are a small family of some thirty species, chiefly of the Southwest Pacific region; only one species is native to the United States.

286b Flowers trimerous **287**

287a (from 286b) Ovary superior; plants mostly climbing by tendrils, not twining, and without prominent tubers .... **Cat-brier Family, SMILACACEAE**

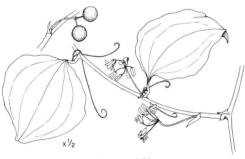

Figure 278

Figure 278 *Smilax rotundifolia* L., Round-leaved Cat-brier, is common in eastern United States.

There are about 300 species of Smilacaceae, widely distributed but mostly in tropical and warm-temperate regions. *Smilax* is by far the largest genus. Many species of *Smilax* have tough, barbed stems, whence the name Cat-brier or Green-brier.

287b Ovary inferior; twining-climbing herbaceous vines, without tendrils, commonly with a large basal tuber .... **Yam Family, DIOSCOREACEAE**

Figure 279

Figure 279 *Dioscorea villosa* L., Wild Yam, grows in open woods thickets, and along roadsides in eastern United States.

This mainly tropical and subtropical family has about 600 species, most of them in the genus *Dioscorea*. Several species of *Dioscorea* are widely cultivated in the tropics for their edible tubers, called Yams, but the "Yams" of grocery stores in the United States are mostly cultivars of *Ipomoea batatas* (L.) Lam. (Sweet Potato), a member of the Convolvulaceae. Species of *Dioscorea* are also important as a source of cortisone.

288a (from 285b) Habit agavoid or yuccoid, i.e., the plants coarse, often shrubby or arborescent xerophytes, with firm or succulent, perennial leaves, the stem very often with secondary growth; stamens six **289**

288b Habit lilioid, i.e., the plants mostly herbs from a bulb or rhizome, with soft, usually annual leaves (leaves perennial but slender and rather lax in *Xerophyllum*, of the Liliaceae); no secondary growth; stamens six or three **290**

289a (from 288a) Leaves firmly succulent; ovules anatropous to sometimes campylotropous .... **Agave Family, AGAVACEAE**

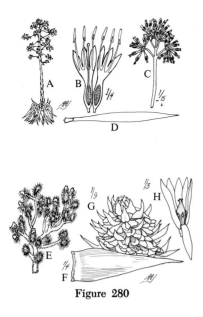

Figure 280

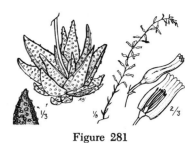

Figure 281

Figure 280 A-D, *Agave americana* L., Century Plant, of deserts in Mexico and southwestern United States, lives for 10-25 years without flowering, then flowers and dies. A, habit; B, flower, in long-section; C, portion of inflorescence; D, leaf. E-H, *Yucca brevifolia* Engelm., Joshua Tree, also of southwestern deserts, is considered to cast about as much shade as a barbed wire fence. E, habit; F, leaf; G, inflorescence; H, flower, with one tepal pulled down.

There are nearly 600 species of Agavaceae, native to warm, mostly arid regions in both the Old World and the New. Some genera, such as *Yucca*, have a superior ovary; others, such as *Agave*, have the ovary inferior. *Yucca* is the state flower of New Mexico.

289b Leaves softly succulent; ovules orthotropous to sometimes hemitropous .... Aloe Family, ALOEACEAE

Figure 281 *Haworthia papillosa* (Salm-Dyck) Haw. is one of several species of *Haworthia* grown indoors in the United States as foliage plants and curiosities.

There are nearly 500 species of Aloeaceae, native chiefly to Africa, Arabia, and Madagascar, especially South Africa. Species of *Aloe*, *Haworthia* and *Gasteria* are familiar as house plants. The Aloeaceae are closely related to the Liliaceae, and have often been included in that family.

290a (from 288b) Perianth woolly; vessels well distributed in all vegetative organs; endosperm fleshy in texture, containing significant amounts of starch as well as other food reserves .... Blood-lily Family, HAEMODORACEAE

Figure 282

Figure 282 *Lachnanthes tinctoria* (Walter) Elliott, Redroot Lily, with red roots and rhizomes, grows in moist, low places in the Atlantic states.

There are only about a hundred species of Haemodoraceae, most of them in the Southern Hemisphere. Except for the wooly flowers they look much like the Liliaceae.

290b Perianth glabrous or sometimes somewhat hairy, but not woolly; vessels chiefly or wholly confined to the roots; endosperm not starchy, often very hard **291**

291a (from 290b) Stamens (including staminodes, if any) mostly six; ovary superior to inferior .... **Lily Family, LILIACEAE**

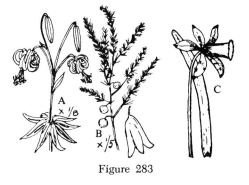

Figure 283

Figure 283 A, *Lilium superbum* L., Turk's-cap Lily, grows in moist places in eastern United States, and similar species occur in the West; B, *Asparagus officinalis* L., Asparagus, has reduced leaves and finely divided green stems (the tender, unexpanded shoot is used as a vegetable in early spring); C, *Narcissus pseudonarcissus* L., Daffodil, with an inferior ovary, represents the segment of Liliaceae that

has often been treated as a separate family Amaryllidaceae. See also fig. 269.

There are nearly 4,000 species of Liliaceae, found throughout most of the world. True Lilies, Day Lilies, Camas, Death Camas, Hyacinth, Amaryllis, Garlic, Onion, *Hosta*, and Tulips are some familiar members of the Liliaceae. The Liliaceae are often regarded as the most "typical" monocotyledons, providing a mental image for the class as a whole. Mayflower, *Maianthemum canadense* Desf. is the state flower of Massachusetts.

291b Stamens three; ovary inferior .... **Iris Family, IRIDACEAE**

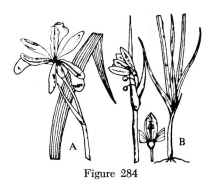

Figure 284

Figure 284 A, *Iris versicolor* L., Blue Flag or Wild Iris, is common in moist, low ground in eastern United States, and the genus is well represented in the West as well as in the East. B, *Sisyrinchium angustifolium* Miller, Blue-eyed Grass, is widespread in the United States.

The Iridaceae are a cosmopolitan family of some 1,500 species, best developed in South Africa. Species of *Crocus, Iris,* and *Gladiolus* are popular garden flowers. Iris is the state flower of Tennessee.

292a (from 276b) Stamen(s) adnate to the style column; staminodes wanting; ovules and seeds very numerous and tiny, the ovules often undeveloped at flowering time; pollen grains usually cohering in pollinia ....
.... Orchid Family, ORCHIDACEAE

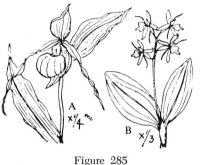

Figure 285

Figure 285   A, *Cypripedium calceolus* L., Yellow Lady-slipper, is widespread but increasingly rare in moist woods in the cooler parts of the Northern Hemisphere; B, *Liparis loeselii* (L.) Rich., Fen Orchis, grows in moist woods in eastern United States.

The Orchidaceae are one of the largest families of plants, with perhaps as many as 20,000 species, the vast majority tropical, many epiphytic, some without chlorophyll. Many species are cultivated under glass for their showy, often bizarre flowers, which are adapted to specific pollinating insects. Only rarely can the "wrong" pollinator effectively transfer pollen from one flower to another. *Cypripedium reginae* Walter, the Showy Lady-slipper, is the state flower of Minnesota.

292b Stamen free from the style; petaloid staminodes generally present; ovules and seeds few to fairly numerous, but of normal size and structure; pollen grains not cohering in pollinia 293

293a (from 292b) Stamen with two pollen-sacs, not petaloid; flowers bilaterally symmetrical; plants aromatic, with abundant oil-cells; leaves and bracts distichously arranged ....
...... Ginger Family, ZINGIBERACEAE

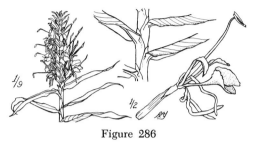

Figure 286

Figure 286   *Hedychium aurantiacum* Wallich, Orange Ginger-Lily, is a native of India, sometimes cultivated under glass in the United States. Ginger is made from the thick, aromatic rhizomes of *Zingiber officinalis* Roscoe, another far-eastern species.

There are about a thousand species of Zingiberaceae, all of warm regions, many of them in southern and southeastern Asia.

293b Stamen with a petaloid blade and a single functional pollen sac, the other sac suppressed; flowers asymmetrical; plants not notably aromatic 294

294a (from 293b) Ovules numerous in each of the three locules of the ovary; leaves spirally arranged; flowers not in mirror-image pairs ....
.... Canna Family, CANNACEAE

the New World. The flowers of *Canna* are asymmetrical. They have three short sepals, three longer, slender petals, four large, petaloid staminodes, and a single functional stamen with a petaloid blade and only one pollen sac.

294b Ovule solitary in the single locule or in each of the three locules of the ovary; leaves distichous; flowers arranged in mirror-image pairs ....
.... **Arrowroot Family, MARANTACEAE**

Figure 287

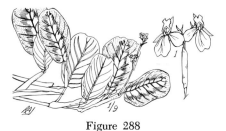

Figure 288

Figure 287   A, *Canna indica* L., Indian Shot, is native to tropical America, rather than to India as Linnaeus supposed. B-E, *XCanna generalis* Bailey, Common Garden Canna, originated in cultivation by hybridization among several species. B, habit; C, flower, front view; D, flower, from behind; E, petaloid stamen, with a single pollen sac at the upper left.

The family has only the genus *Canna*, with about fifty species, native to warm regions in

Figure 288   *Maranta leuconeura* E. Morren is grown indoors in United States for its ornamental foliage, as are several other species of *Maranta* and *Calathea*.

There are about 350 species of Marantaceae, all tropical or subtropical, mostly of moist or wet, often shady places, especially in the New World.

Outline of Classification of the Living Families of Seed Plants

Families treated in the text are in Roman type; those not treated are in italics.

I. DIVISION PINOPHYTA
The Gymnosperms
A. SUBDIVISION CYCADICAE
1. Class Cycadopsida
a. Order Cycadales
1. Family Cycadaceae

B. SUBDIVISION PINICAE
1. Class Ginkgoopsida
a. Order Ginkgoales
1. Family Ginkgoaceae
2. Class Pinopsida
a. Order Pinales, the Conifers
1. Family Pinaceae
2. Family Taxodiaceae
3. Family Cupressaceae
4. Family Araucariaceae
5. Family *Podocarpaceae*
6. Family *Cephalotaxaceae*
b. Order Taxales
1. Family Taxaceae

C. SUBDIVISION GNETICAE
1. Class Ephedropsida
a. Order Ephedrales
1. Family Ephedraceae
2. Class Welwitschiopsida
a. Order Welwitschiales
1. Family *Welwitschiaceae*

3. Class Gnetopsida
a. Order Gnetales
1. Family Gnetaceae

II. DIVISION MAGNOLIOPHYTA
The Angiosperms
A. CLASS MAGNOLIOPSIDA, THE DICOTYLEDONS
1. Subclass Magnoliidae
a. Order Magnoliales
1. Family Winteraceae
2. Family *Degeneriaceae*
3. Family *Himantandraceae*
4. Family Magnoliaceae
5. Family *Lactoridaceae*
6. Family *Austrobaileyaceae*
7. Family *Eupomatiaceae*
8. Family Annonaceae
9. Family Myristicaceae
10. Family Canellaceae
b. Order Laurales
1. Family *Amborellaceae*
2. Family *Trimeniaceae*
3. Family *Monimiaceae*
4. Family *Gomortegaceae*

5. Family Calycanthaceae
6. Family *Idiospermaceae*
7. Family Lauraceae
8. Family *Hernandiaceae*
c. Order Piperales
1. Family *Chloranthaceae*
2. Family Saururaceae
3. Family Piperaceae
d. Order Aristolochiales
1. Family Aristolochiaceae
e. Order Illiciales
1. Family Illiciaceae
2. Family Schisandraceae
f. Order Nymphaeales
1. Family Nelumbonaceae
2. Family Nymphaeaceae
3. Family Cabombaceae
4. Family Ceratophyllaceae
g. Order Ranunculales
1. Family Ranunculaceae
2. Family *Circaeasteraceae*
3. Family Berberidaceae
4. Family *Sargentodoxaceae*
5. Family Lardizabalaceae
6. Family Menispermaceae
7. Family Coriariaceae
8. Family Corynocarpaceae
9. Family *Sabiaceae*

h. Order Papaverales
  1. Family Papaveraceae
  2. Family Fumariaceae

2. Subclass Hamamelidae
a. Order Trochoden-
   drales
  1. Family *Tetracentraceae*
  2. Family *Trochodendra-*
   ceae

b. Order Hamameli-
   dales
  1. Family Cercidiphyllaceae
  2. Family *Eupteliaceae*
  3. Family Platanaceae
  4. Family Hamamelida-
   ceae
  5. Family *Myrothamna-*
   ceae

c. Order Didymelales
  1. Family *Didymelaceae*

d. Order Eucommiales
  1. Family *Eucommiaceae*

e. Order Urticales
  1. Family *Barbeyaceae*
  2. Family Ulmaceae
  3. Family Moraceae
  4. Family Cannabaceae
  5. Family Urticaceae

f. Order Leitneriales
  1. Family Leitneriaceae

g. Order Juglandales
  1. Family *Rhoipteleaceae*
  2. Family Juglandaceae

h. Order Myricales
  1. Family Myricaceae

i. Order Fagales
  1. Family *Balanopaceae*
  2. Family Fagaceae
  3. Family Betulaceae

j. Order Casuarinales
  1. Family Casuarinaceae

3. Subclass Caryophyl-
   lidae
a. Order Caryophyl-
   lales
  1. Family Phytolaccaceae
  2. Family *Achatocarpa-*
   ceae
  3. Family Nyctaginaceae
  4. Family Aizoaceae
  5. Family *Didiereaceae*
  6. Family Cactaceae

  7. Family Chenopodia-
   ceae
  8. Family Amaranthaceae
  9. Family Portulacaceae
  10. Family Basellaceae
  11. Family Molluginaceae
  12. Family Caryophylla-
   ceae

b. Order Polygonales
  1. Family Polygonaceae

c. Order Plumba-
   ginales
  1. Family Plumbagina-
   ceae

4. Subclass Dilleniidae
a. Order Dilleniales
  1. Family Dilleniaceae
  2. Family Paeoniaceae
  3. Family Crossosomata-
   ceae

b. Order Theales
  1. Family Ochnaceae
  2. Family *Sphaerosepala-*
   ceae
  3. Family *Sarcolaenaceae*
  4. Family Dipterocarpa-
   ceae
  5. Family *Stachyuraceae*
  6. Family *Caryocaraceae*
  7. Family Theaceae
  8. Family *Paracryphiaceae*
  9. Family Actinidiaceae
  10. Family *Pentaphylaca-*
   ceae
  11. Family *Oncothecaceae*
  12. Family *Tetramerista-*
   ceae
  13. Family *Pellicieraceae*
  14. Family *Marcgraviaceae*
  15. Family *Quiinaceae*
  16. Family Elatinaceae
  17. Family *Medusagyna-*
   ceae
  18. Family Clusiaceae

c. Order Malvales
  1. Family *Elaeocarpaceae*
  2. Family *Scytopetalaceae*
  3. Family *Huaceae*
  4. Family Tiliaceae
  5. Family Sterculiaceae
  6. Family Bombacaceae
  7. Family Malvaceae

d. Order Lecythidales
  1. Family Lecythidaceae

e. Order Nepenthales
  1. Family Sarraceniaceae

  2. Family Nepenthaceae
  3. Family Droseraceae

f. Order Violales
  1. Family Flacourtiaceae
  2. Family *Peridiscaceae*
  3. Family Bixaceae
  4. Family Cistaceae
  5. Family *Lacistemaceae*
  6. Family *Scyphostegia-*
   ceae
  7. Family Violaceae
  8. Family Tamaricaceae
  9. Family *Frankeniaceae*
  10. Family *Dioncophylla-*
   ceae
  11. Family *Ancistroclada-*
   ceae
  12. Family Turneraceae
  13. Family *Malesherbia-*
   ceae
  14. Family Passifloraceae
  15. Family *Fouquieriaceae*
  16. Family *Hoplestigmata-*
   ceae
  17. Family Achariaceae
  18. Family Caricaceae
  19. Family Cucurbitaceae
  20. Family Datiscaceae
  21. Family Begoniaceae
  22. Family Loasaceae

g. Order Salicales
  1. Family Salicaceae

h. Order Capparales
  1. Family *Tovariaceae*
  2. Family Capparaceae
  3. Family Brassicaceae
  4. Family Moringaceae
  5. Family Resedaceae

i. Order Batales
  1. Family *Gyrostemona-*
   ceae
  2. Family Bataceae

j. Order Ericales
  1. Family Cyrillaceae
  2. Family Clethraceae
  3. Family *Grubbiaceae*
  4. Family Empetraceae
  5. Family *Epacridaceae*
  6. Family Ericaceae
  7. Family Pyrolaceae
  8. Family Monotropaceae

k. Order Diapensiales
  1. Family Diapensiaceae

l. Order Ebenales
  1. Family Sapotaceae

2. Family Ebenaceae
3. Family Styracaceae
4. Family *Lissocarpaceae*
5. Family Symplocaceae

m. Order Primulales
1. Family Theophrastaceae
2. Family Myrsinaceae
3. Family Primulaceae

5. Subclass Rosidae
a. Order Rosales
1. Family *Brunelliaceae*
2. Family Connaraceae
3. Family *Eucryphiaceae*
4. Family Cunoniaceae
5. Family *Davidsoniaceae*
6. Family *Dialypetalanthaceae*
7. Family Pittosporaceae
8. Family *Byblidaceae*
9. Family Hydrangeaceae
10. Family Columelliaceae
11. Family Grossulariaceae
12. Family *Greyiaceae*
13. Family *Bruniaceae*
14. Family *Anisophylleaceae*
15. Family *Alseuosmiaceae*
16. Family Crassulaceae
17. Family Cephalotaceae
18. Family Saxifragaceae
19. Family Donatiaceae
20. Family Rosaceae
21. Family Neuradaceae
22. Family Chrysobalanaceae
23. Family Surianaceae
24. Family *Rhabdodendraceae*

b. Order Fabales
1. Family Mimosaceae
2. Family Caesalpiniaceae
3. Family Fabaceae

c. Order Podostemales
1. Family Podostemaceae

d. Order Haloragales
1. Family Haloragaceae
2. Family Gunneraceae

e. Order Myrtales
1. Family *Sonneratiaceae*
2. Family Lythraceae
3. Family *Penaeaceae*
4. Family *Crypteroniaceae*
5. Family Thymelaeaceae
6. Family Trapaceae
7. Family Myrtaceae

8. Family Punicaceae
9. Family Onagraceae
10. Family *Oliniaceae*
11. Family Melastomataceae
12. Family Combretaceae

f. Order Proteales
1. Family Elaeagnaceae
2. Family Proteaceae

g. Order Rhizophorales
1. Family Rhizophoraceae

h. Order Cornales
1. Family *Alangiaceae*
2. Family Nyssaceae
3. Family Cornaceae
4. Family Garryaceae

i. Order Santalales
1. Family *Medusandraceae*
2. Family *Dipentodontaceae*
3. Family Olacaceae
4. Family *Opiliaceae*
5. Family Santalaceae
6. Family *Misodendraceae*
7. Family Loranthaceae
8. Family Viscaceae
9. Family *Eremolepidaceae*
10. Family *Balanophoraceae*

j. Order Rafflesiales
1. Family Hydnoraceae
2. Family *Mitrastemonaceae*
3. Family Rafflesiaceae

k. Order Celastrales
1. Family *Geissolomataceae*
2. Family Celastraceae
3. Family *Hippocrateaceae*
4. Family *Stackhousiaceae*
5. Family *Salvadoraceae*
6. Family Aquifoliaceae
7. Family Icacinaceae
8. Family *Aextoxicaceae*
9. Family *Cardiopteridaceae*
10. Family *Dichapetalaceae*

l. Order Euphorbiales
1. Family Buxaceae
2. Family Simmondsiaceae
3. Family *Daphniphyllaceae*

4. Family *Pandaceae*
5. Family Euphorbiaceae

m. Order Rhamnales
1. Family Rhamnaceae
2. Family Leeaceae
3. Family Vitaceae

n. Order Linales
1. Family Erythroxylaceae
2. Family *Humiriaceae*
3. Family *Ixonanthaceae*
4. Family *Hugoniaceae*
5. Family Linaceae

o. Order Polygalales
1. Family Malpighiaceae
2. Family *Vochysiaceae*
3. Family *Trigoniaceae*
4. Family *Tremandraceae*
5. Family Polygalaceae
6. Family *Xanthophyllaceae*
7. Family Krameriaceae

p. Order Sapindales
1. Family Staphyleaceae
2. Family *Melianthaceae*
3. Family *Bretschneideraceae*
4. Family *Akaniaceae*
5. Family Sapindaceae
6. Family Hippocastanaceae
7. Family Aceraceae
8. Family Burseraceae
9. Family Anacardiaceae
10. Family *Julianiaceae*
11. Family Simaroubaceae
12. Family Cneoraceae
13. Family Meliaceae
14. Family Rutaceae
15. Family Zygophyllaceae

q. Order Geraniales
1. Family Oxalidaceae
2. Family Geraniaceae
3. Family Limnanthaceae
4. Family Tropaeolaceae
5. Family Balsaminaceae

r. Order Apiales
1. Family Araliaceae
2. Family Apiaceae

6. Subclass Asteridae
a. Order Gentianales
1. Family Loganiaceae
2. Family Gentianaceae
3. Family *Saccifoliaceae*
4. Family Apocynaceae
5. Family Asclepiadaceae

b. Order Solanales
1. Family *Duckeodendra-ceae*
2. Family *Nolanaceae*
3. Family Solanaceae
4. Family Convolvulaceae
5. Family Cuscutaceae
6. Family Menyanthaceae
7. Family *Retziaceae*
8. Family Polemoniaceae
9. Family Hydrophylla-ceae

c. Order Lamiales
1. Family Lennoaceae
2. Family Boraginaceae
3. Family Verbenaceae
4. Family Lamiaceae

d. Order Callitrichales
1. Family Hippuridaceae
2. Family Callitrichaceae
3. Family *Hydrostachya-ceae*

e. Order Plantaginales
1. Family Plantaginaceae

f. Order Scrophulariales
1. Family Buddlejaceae
2. Family Oleaceae
3. Family Scrophularia-ceae
4. Family *Globulariaceae*
5. Family *Myoporaceae*
6. Family Orobanchaceae
7. Family Gesneriaceae
8. Family Acanthaceae
9. Family Pedaliaceae
10. Family Bignoniaceae
11. Family *Mendonciaceae*
12. Family Lentibularia-ceae

g. Order Campanulales
1. Family Pentaphragma-taceae
2. Family *Sphenocleaceae*
3. Family Campanulaceae
4. Family *Stylidiaceae*
5. Family *Brunoniaceae*
6. Family Goodeniaceae

h. Order Rubiales
1. Family Rubiaceae
2. Family *Theligonaceae*

i. Order Dipsacales
1. Family Caprifoliaceae
2. Family Adoxaceae
3. Family Valerianaceae

4. Family Dipsacaceae
5. Family Calyceraceae

j. Order Asterales
1. Family Asteraceae

B. CLASS LILIOPSIDA, THE MONOCOTYLEDONS
1. Subclass Alismatidae
a. Order Alismatales
1. Family Butomaceae
2. Family Limnocharita-ceae
3. Family Alismataceae

b. Order Hydrochari-tales
1. Family Hydrocharita-ceae

c. Order Najadales
1. Family *Aponogetona-ceae*
2. Family Scheuchzeria-ceae
3. Family Juncaginaceae
4. Family Potamogetona-ceae
5. Family Ruppiaceae
6. Family Zosteraceae
7. Family Najadaceae
8. Family Zannichellia-ceae
9. Family Cymodoceaceae

d. Order Triuridales
1. Family *Petrosaviaceae*
2. Family *Triuridaceae*

2. Subclass Arecidae
a. Order Arecales
1. Family Arecaceae

b. Order Cyclanthales
1. Family Cyclanthaceae

c. Order Pandanales
1. Family Pandanaceae

d. Order Arales
1. Family Araceae
2. Family Lemnaceae

3. Subclass Commelinidae
a. Order Commelinales
1. Family *Rapateaceae*
2. Family Xyridaceae
3. Family Mayacaceae
4. Family Commelinaceae

b. Order Eriocaulales

1. Family Eriocaulaceae

c. Order Restionales
1. Family *Flagellariaceae*
2. Family *Joinvilleaceae*
3. Family Restionaceae
4. Family *Centrolepida-ceae*
5. Family *Hydatellaceae*

d. Order Juncales
1. Family Juncaceae
2. Family *Thurniaceae*

e. Order Cyperales
1. Family Cyperaceae
2. Family Poaceae

f. Order Typhales
1. Family Sparganiaceae
2. Family Typhaceae

4. Subclass Zingiberidae
a. Order Bromeliales
1. Family Bromeliaceae

b. Order Zingiberales
1. Family Strelitziaceae
2. Family Heliconiaceae
3. Family Musaceae
4. Family *Lowiaceae*
5. Family Zingiberaceae
6. Family Costaceae
7. Family Cannaceae
8. Family Marantaceae

5. Subclass Liliidae
a. Order Liliales
1. Family *Philydraceae*
2. Family Pontederiaceae
3. Family Haemodoraceae
4. Family *Cyanastraceae*
5. Family Liliaceae
6. Family Iridaceae
7. Family *Velloziaceae*
8. Family Aloeaceae
9. Family Agavaceae
10. Family *Xanthorrhoea-ceae*
11. Family *Hanguanaceae*
12. Family *Taccaceae*
13. Family Stemonaceae
14. Family Smilacaceae
15. Family Dioscoreaceae

b. Order Orchidales
1. Family *Geosiridaceae*
2. Family Burmanniaceae
2. Family *Corsiaceae*
4. Family Orchidaceae

Index and
Pictured Glossary

The definitions in the glossary are for usage in the seed plants. Many of the terms also have a broader meaning, or a different meaning in some other groups.

A

A- AB-: Latin prefix meaning not, or different from, or away from, or without.

ABAXIAL: Away from the axis.

Abelia, 114

Abies, 14

Abronia, 71

Acacia, 50

Acanthaceae, 108

Acanthus family, 108

ACCRESCENT: Increasing in size with age, as for example a calyx that continues to enlarge while the fruit is maturing.

Acer, 3, 4, 43

Aceraceae, 43

ACHENE: The most generalized type of dry, indehiscent fruit, lacking the specialized features that mark, for example, a samara or nut.

Achlys, 45

Actaea, 52

Actinidia, 54, 55

Actinidiaceae, 54

AD-: Latin prefix meaning to or toward.

ADAXIAL: Toward the axis.

Adansonia, 55

ADNATE: Grown together, or attached. The term is applied only to unlike parts, as stipules adnate to the petiole, or stamens adnate to the corolla. (Compare connate.)

Adonis, 28

Adoxa, 97

Adoxa family, 97

Adoxaceae, 97

Aesculus, 79

African Tulip Tree, 109

African Violet, 109

Agavaceae, 131

Agave, 132

Agave family, 131

Agrimonia, 30

Agrimony, 30

Ailanthus, 62

Aizoaceae, 52, 84

Akebia, 21

Alabama state flower, 54

Alaska state flower, 100

Albizia, 96

Alder, 40

Alfalfa, 81

Alisma, 117

Alismataceae, 117

Allamanda, 103

Alleghany Spurge, 42

Almond, 91

Alnus, 40

Aloe family, 132

ALTERNATE: Placed singly at each node, as the leaves on a stem; placed regularly between organs of another kind, as stamens alternate with the petals.

Figure 289

Althaea, 56

Alum Root, 75

Aluminum Plant, 71

Amaranth family, 72

Amaranthaceae, 72

Amaranthus, 72

Amaryllidaceae, 133

Amaryllis, 133

Ambrosia, 112

Amelanchier, 87

AMENT: A dense, bracteate spike or raceme with a non-fleshy axis bearing many small, naked or apetalous flowers; a catkin. (Compare spadix.) See Fig. 297.

American Bittersweet, 68

Ammobroma, 20

Amoreuxia, 46

Anacardiaceae, 48, 64

Ananas, 128

ANATROPOUS OVULE: An ovule that is bent back on itself, so that the micropyle is next to the funiculus. See Fig. 315.

ANDROGYNOPHORE: A stalk, arising from the receptacle of a flower, on which both the stamens and the pistil of some kinds of flowers are borne.

Andromeda, 96

Anemone, 28

Anemopsis, 91

Anethole, 26

Angiosperms, 7, 10, 15

Angle-Pod, 103

Anise, 90

Anise oil, 26

Annonaceae, 25

ANTHER: The part of a stamen, consisting of one or usually two pollen sacs (and a connecting layer between them), that bears the pollen.

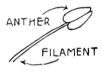

Figure 290

ANTHESIS: The stage at which a flower is fully developed, shedding or receptive to pollen; full bloom.

Anthurium, 123

Apacheria, 24

APETALOUS: Without petals.

Aphelandra, 108

Apiaceae, 90

Apocynaceae, 103

Apocynum, 103

a posteriori, 6

APOTROPOUS OVULE: An ovule which (if erect) has the raphe ventral, (between the partition and the body of the ovule), or which (if pendulous) has the raphe dorsal. (Compare epitropous ovule.)

Apple, 87

a priori, 6

aquatic plants, 15, 26, 30, 104, 115

Aquifoliaceae, 68

Aquilegia, 28

Araceae, 123

Aralia, 89

Araliaceae, 89

Araucaria, 13

Araucaria family, 13

Araucariaceae, 13

ARBORESCENT: Tree-like, or becoming a tree, or almost a tree.

Arceuthobium, 18

Ardisia, 95

Arecaceae, 122

Argemone, 46
ARIL: A specialized outgrowth from the funiculus that covers or is attached to the mature seed; more loosely, any appendage or fleshy thickening of the seed coat.
ARILLATE: Provided with an aril.
Arisaema, 123
Aristolochia, 85
Aristolochiaceae, 57, 85
Arizona state flower, 84
Armeria, 94
Aroid family, 123
Arrowgrass, 121
Arrowgrass family, 121
Arrowhead, 117
Arrowroot family, 135
Artemisia, 112
Artichoke, 112
Artocarpus, 36
Asarum, 85
Asclepiadaceae, 103
Asclepias, 103
Ash, 36
Asimina, 26
Asparagus, 133
Aster, 112
Aster family, 111
Asteraceae, 111
Astragalus, 81
Astrophytum, 84
Atriplex, 65
Australian Brush-Cherry, 86
Australian Pine family, 36
AUTOTROPH: A plant that is nutritionally independent, making its own food from raw materials obtained more or less directly from the substrate. The term is usually interpreted to include mycorhizal as well as nonmycorhizal plants, so long as they are photosynthetic.
Avocado, 44
AXIL: The point of the angle formed by the leaf or petiole with the upward internode of the stem.
AXILE PLACENTA: A placenta along the central axis (or along the vertical midline of the partition) of an ovary with two or more locules.

Figure 291

AXILLARY: Located in or arising from an axil.
Azalea, 96, 101

B

Baby's Breath, 49
Baby's Tears, 71
Bald Cypress family, 14
Balsa wood, 55
Balsam, 78
Balsaminaceae, 77
Banana family, 129
Banisteriopsis, 79
Baobab Tree, 55
Barbados Cherry, 79
Barbados Cherry family, 79
Barberry family, 44, 52
BASAL PLACENTA: A placenta at the base of an ovary, usually in a unilocular ovary.

Figure 292

Basella family, 70
Basellaceae, 70
Basswood, 56
Basswood family, 56
Bataceae, 37
Batis, 37
Batis family, 37
Bay Cedar family, 24
Bayberry family, 39
Beans, 81
Beard-tongue, 110
Beautyberry, 100
Bedstraw, 113
Bee-plant, 69
Beech, 40
Beech family, 40
Beets, 73
Begonia, 86
Begonia family, 85
Begoniaceae, 85
Beloperone, 108
Ben Oil, 83
Berberidaceae, 44, 52
Berberis, 45
Bergia, 33
BERRY: The most generalized type of fleshy fruit, developed from a single pistil, fleshy throughout, and containing usually several or many seeds; more loosely, any pulpy or juicy fruit.

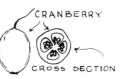

Figure 293

Bertholletia, 87
Besseya, 8
Beta, 73
Betula, 40
Betulaceae, 39
Bignoniaceae, 108
BI-: Latin prefix meaning two.
BIFID: More or less deeply cleft from the tip into two usually equal parts.
Billbergia, 128
BILOBED: With 2 lobes.
BILOCULAR: Having or composed of two locules.
Bindweed, 106
binomial system, 3, 4
BIPINNATE: Twice pinnate.
Birch, 40
Birch family, 39
Bird-of-Paradise Flower family, 129
Birthwort family, 57, 85
BISERIATE: Arranged in two series.
Bishop's Cap, 84
Bittersweet family, 67
Bixa, 46
Bixaceae, 46
Black-eyed Susan, 112
Black Gum, 92
Black Walnut, 35
Blackberry, 30
Bladdernut, 67
Bladdernut family, 66
Bladderwort, 15
Bladderwort family, 15
Blazing Star, 112
Bleeding Heart, 82
Blood-lily family, 132
Bloodroot, 45
Blue-eyed Grass, 133
Blue Flag, 133
Blue Gum, 86
Blue-weed, 100
Bluebell, 100
Bluebell of Scotland, 113
Blueberry, 96
Bluebonnet, 81
Bluegrass, 126
Bluets, 113
Boehmeria, 71
Bombacaceae, 55
Borage family, 99
Boraginaceae, 6, 99
BOREAL: Northern; in the northern part of the Northern Hemisphere.
Boswellia, 62
Bougainvillea, 52
Boussingaultia, 70
Box family, 41
Boxwood, 42
BRACT: A specialized leaf, from the axil of which a flower or flower stalk arises; more loosely, any more or less reduced or modified leaf associated with a flower or an inflorescence, but not a part of the flower itself; in conifers, one of the primary appendages of the cone axis, in the axils of which the ovule-bearing cone-scales arise.

Figure 294

BRACTEATE: Provided with bracts.
Brasenia, 27
Brassica, 69
Brassicaceae, 69
Brazil Nut, 87
Brazil Nut family, 87
Breadfruit, 36
Bridal Wreath, 23
Broccoli, 69
Bromeliaceae, 128
Broom-Rape family, 19
Brussels Sprouts, 69
Buckbean, 105
Buckbean family, 104
Buckthorn, 63
Buckthorn family, 63
Buckwheat, 50
Buckwheat family, 33, 49, 73
Buddleja, 104
Buddlejaceae, 104
Buffalo Berry, 64
Bulrush, 126
Bumelia, 98
Bur-reed family, 121
Burmannia, 130
Burmannia family, 130
Burmanniaceae, 130
Burning Bush, 68
Bursera, 62
Burseraceae, 61
Butomaceae, 117
Butomus, 118
Butter and Eggs, 110
Buttercup, 3, 28
Buttercup family, 28, 52
Butterfly Bush, 104
Butterfly Bush family, 104
Butterfly-Weed, 103
Butterwort, 15
Buxaceae, 41
Buxus, 42

C

Cabbage, 69
Cabomba, 27
Cabomba family, 27
Cabombaceae, 27, 32
Cacao family, 56, 63
Cactaceae, 83, 93
Cactus family, 83, 93
Caesalpinia family, 81
Caesalpiniaceae, 81
caffeine, 56
Caladium, 123
Calathea, 135
Calceolaria, 110
California Big Tree, 15
California Laurel, 44
California Poppy, 45
California state flower, 45
Callicarpa, 100
Callitrichaceae, 34

Callitriche, 34
Calluna, 96
Calochortus, 127
Caltrop, 60
Calycanthaceae, 23
Calycanthus, 24
CALYX: All the sepals of a flower, collectively.
Camas, 133
Camellia, 54
Campanula, 113
Campanulaceae, 113
Camphor, 44
Campsis, 109
Camptotheca, 92
CAMPYLOTROPOUS OVULE: An ovule distorted by unequal growth, so that the body seems to lie centrally transverse to the stalk.
Canacomyrica, 39
Cancer-Root, 19
Canella, 47
Canella family, 47
Canellaceae, 47
Canna, 135
Canna family, 134
Cannabaceae, 71
Cannabis, 6, 72
Cannaceae, 134
Caper-Bush, 47
Caper family, 47, 69, 82
CAPITATE: Headlike, or in a head.
Capparaceae, 47, 69, 82
Capparis, 47
Caprifoliaceae, 114
Capsella, 4
CAPSULE: A dry, dehiscent fruit composed of more than one carpel.

Figure 295

Caraway, 90
Cardinal Flower, 113
Carex, 126
Carica, 98
Caricaceae, 98
Carludovica, 122
Carnation, 49
Carnegiea, 84
CARPEL: The fertile leaf (or serial homologue of a leaf) of an angiosperm, which bears the ovules. One or more carpels join to compose a pistil.

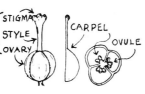

Figure 296

Carpet Weed, 76
Carpet-Weed family, 76
Carpinus, 40
CARPOPHORE: The part of the receptacle which in some kinds of flowers is prolonged between the carpels as a central axis, e.g., in the Apiaceae.
Carrot, 90
Carrot family, 90
Carya, 3, 35
Caryophyllaceae, 49, 73
Cassava, 41
Cassia, 81
Cassytha, 17
Castanea, 40
Castilleja, 110
Castor Bean, 41
Casuarina, 37
Casuarinaceae, 36
Cat-brier family, 131
Catalpa, 109
Catclaw Acacia, 50
CATKIN: A dense, bracteate spike or raceme with a non-fleshy axis bearing many small, naked or apetalous flowers. An ament. (Compare spadix.)

Figure 297

Catnip, 101
Cattail, 121
Cattail family, 121
Cauliflower, 69
Ceanothus, 63
Cecropia, 36
Cedrela, 66
Ceiba, 55
Celandine, 46
Celastraceae, 67
Celastrus, 68
Celery, 90
Celosia, 72
Celtis, 66
Century Plant, 132
Cerastium, 49
Ceratiola, 67
Ceratophyllaceae, 32
Ceratophyllum, 32
Cercidiphyllaceae, 22
Cercidiphyllum, 23
Cercidiphyllum family, 22
Cercis, 81

Cereals, 126
Ceropegia, 103
Chamaecyparis, 14
chaparral, 63
characters, taxonomic value of, 6
Chaulmoogra oil, 47
Chaulmoogra Tree, 47
Cheeses, 56
Chelidonium, 46
Chenopodiaceae, 65, 72
Chenopodium, 73
Chestnut, 40
Chickweed, 49
chicle, 98
China-Berry, 62
Chinese Gooseberry, 55
Chinese Gooseberry family, 54
Chinese Lantern Plant, 107
Chionanthus, 107
CHLOROPHYLL: The characteristic green pigment of plants, an essential enzyme in photosynthesis.
chocolate, 56
Christmas Cactus, 94
Christmas Rose, 28
Chrysanthemum, 112
Chrysobalanaceae, 50
Chrysobalanus, 50
Chrysobalanus family, 50
Chrysophyllum, 98
Cinchona, 113
Cinnamomum, 44
cinnamon, 44
CIRCUM-: Latin prefix meaning around, as around a circle.
CIRCUMBOREAL: Occurring in northern regions all around the world.
CIRCUMSCISSILE: Dehiscing by an encircling transverse line, so that the top comes off as a lid or cap.
Cistaceae, 48
Citronella, 65
Citrus family, 42
Clarkia, 89
class, 2
classification, 1, 6
  categories of, 2
  of Seeds Plants, 6
  outline of, 136
Claytonia, 70
Clematis, 20, 28
Cleome, 69, 82
Clethra, 60
Clethraceae, 60
CLIMAX COMMUNITY: A potentially stable community, capable of maintaining itself in dynamic equilibrium as long as the climate remains unchanged and there is no catastrophic disturbance. (Compare seral.)
Clover, 58, 81
Clusiaceae, 42
Cobra Plant, 16, 17
Coca family, 60
Coca Plant, 61
cocaine, 61
Cocculus, 22
Cochlospermum, 46
Cockscomb, 72
Coco Plum, 50
cocoa, 56

Coffea, 113
coffee, 113
Cola, 56
colanine, 56
Coleus, 101
Columbine, 28
Comandra, 91
Combretaceae, 91
Combretum, 91
Commelina, 128
Commelinaceae, 128
Commiphora, 62
competitive exclusion, 4
COMPOUND LEAF: A leaf with two or more distinct leaflets. (See palmately compound; pinnately compound; trifoliolate.)
COMPOUND OVARY OR PISTIL: An ovary or pistil composed of more than one carpel.
COMPOUND UMBEL: An umbel whose primary branches are themselves umbellately branched. See Fig. 336.
Comptonia, 39
CONNATE: Grown together or attached. The term is applied only to organs of the same kind, as filaments connate into a tube, or leaves connate around the stem. (Compare adnate.)
Connecticut state flower, 96
CONNECTIVE: The tissue connecting the two pollen sacs of an anther.
CONVOLUTE: Arranged so that each member of the set has one edge exposed (covering the adjoining member) and the other edge covered, no one member being wholly external or wholly internal to the others.
Convolvulaceae, 106
Convolvulus, 106
Coralberry, 114
CORDATE: Shaped like a stylized heart, with a more or less pointed tip and rounded to a notch at the base.
Corema, 67
Coriander, 90
Cork-tree, 43
Corkwood family, 39
Cornaceae, 92
Cornel, 93
Cornus, 6, 93
COROLLA: All the petals of a flower, collectively.
COROLLOID: Resembling a corolla in some way.
CORONA: A set of petal-like structures or appendages between the corolla and the stamens, derived by modification of the corolla or of some of the stamens.
cortisone, 131
Corydalis, 82
Corylus, 39, 40

cotton, 56
COTYLEDON: A leaf of the embryo of a seed.

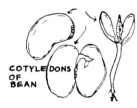

COTYLEDONS OF BEAN

Figure 298

Cranberry, 96
Crape Myrtle, 54
Crassula, 29
Crassulaceae, 28
Crataegus, 87
Creosote, 60
Creosote Bush family, 59
Crocus, 133
Croomia, 131
Crossosoma, 24
Crossosoma family, 24
Crossosomataceae, 24
Croton, 41
Crowberry, 67
Crowberry family, 67
Cruciferae, 69
Cucumber, 111
Cucumis, 111
Cucurbitaceae, 111
Cumin, 90
Cupressaceae, 14
Cupressus, 14
CUPULATE: Cup-shaped.
CUPULE: Diminutive of cup; a small structure shaped like a cup.
curare, 22, 104
Currants, 88
Cuscuta, 17
Cuscutaceae, 17
Custard-Apple family, 25
Cycad family, 10
Cycadaceae, 10
Cycas, 10
Cyclanthaceae, 122
Cyclanthus family, 122
CYME: A broad class of inflorescences characterized by having the terminal flower bloom first, commonly also with the terminal flower of each branch blooming before the others on that branch. (Compare raceme, racemose.)

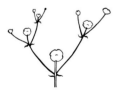

Figure 299

Cymodoceaceae, 118
Cynara, 112
Cynoglossum, 100
Cyperaceae, 125
Cyperus, 126
Cypress family, 14
Cypress Vine, 106
Cypripedium, 134
Cyrilla, 68
Cyrilla family, 68
Cyrillaceae, 68
CYSTOLITH: A relatively large, intracellular concretion of calcium carbonate, silica, or other minerals, found for example in the epidermal cells of the leaves of Cannabaceae, Moraceae, and Urticaceae.

D

Daffodil, 133
Dahlia, 112
Daisy, 112
Dandelion, 112
Daphne, 51
Daphne family, 51
Darlingtonia, 16
Datisca, 88
Datisca family, 88
Datiscaceae, 88
Datura, 107
Davidia, 92
Dawn Redwood, 15
Day-flower, 128
Day Lilies, 133
Death Camas, 133
DECIDUOUS: Falling after completion of the normal function. A deciduous tree is one that normally loses its leaves at the approach of winter or the dormant season.
DECUSSATE: Arranged oppositely, with each succeeding pair set at right angles to the previous pair.

Figure 300

DEHISCENT: Opening at maturity, releasing or exposing the contents.
Delonix, 81
Delphinium, 28
Deutzia, 87
Dianthus, 49
Diapensia, 102
Diapensia family, 102
Diapensiaceae, 102

DI-: Greek prefix meaning two.
Dicentra, 82
DICHLAMYDEOUS: Having two series or sets of perianth members, ordinarily sepals and petals (Compare monochlamydeous.)
DICHOTOMOUS: Forking more or less regularly into two branches of about equal size.
DICOTYLEDONS: A major group of angiosperms, in which the embryo typically has two cotyledons.
Dicotyledons, 15
Dictamnus, 43
Didiplis, 32
Dieffenbachia, 123
Dill, 90
Dillenia family, 25
Dilleniaceae, 25
DIMEROUS: With parts in sets of two.
DIOECIOUS: Producing male and female flowers (or other reproductive structures) on separate individuals.
dioecious, 12
Dionaea, 16
Dioscorea, 131
Dioscoreaceae, 131
Diospyros, 98
Dipsacaceae, 114
Dipsacus, 114
Dipterocarp family, 61
Dipterocarpaceae, 61
Dipterocarpus, 61
Dirca, 51
DISK, OR NECTARY DISK: An outgrowth from the receptacle or from the hypanthium, surrounding the base of the ovary, often derived by evolutionary reduction of the innermost set of stamens, and commonly producing nectar; in the Asteraceae the central part of the head, composed of tubular flowers, is called the disk.
DISSECTED: Deeply (and often repeatedly) divided into numerous small or slender parts.
DISTAL: At the far end. (Compare proximal.)
DISTICHOUS: In two vertical rows or ranks, on opposite sides of an axis.
DISTINCT: Not connate with similar organs.
Ditch Grass, 120
Ditch-Grass family, 120
division, 2
Dodder, 17
Dodder family, 17
Dodecatheon, 95
Dogbane, 103
Dogbane family, 103
Dogwood, 93
Dogwood family, 92
Doll's Eyes, 52
Dombeya, 56
DORSAL: Pertaining to or located on the back; on the

side of an appendage that is away from the axis. (Compare ventral.)
Douglas Fir, 14
Drimys, 26
Drimys family, 26
Drosera, 16
Droseraceae, 16
DRUPACEOUS: Consisting of a drupe, or like a drupe.
DRUPE: A fleshy fruit with a central stone enclosing the seed (or with 2-several such stones each enclosing a seed).

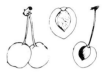

Figure 301

Duckweed, 115
Duckweed family, 115
Dutchman's Breeches, 82

E

E- EX-: Latin prefix meaning without, or from, or away from.
Eastern Red Cedar, 14
Ebenaceae, 97
Ebony family, 97
Ebony wood, 98
Echinocactus, 84
Echium, 100
ecological niche, 4
Eel Grass, 119
Eel Grass family, 119
Eichhornia, 117
Elaeagnaceae, 64
Elaeagnus, 64
Elatinaceae, 33, 76
Elatine, 33
Elderberry, 114
Ellisia, 106
Elm, 66
Elm family, 65
Elodia, 116
EMBRYO: The young plant in a seed.
EMBRYO SAC: The central portion of an ovule, within which the embryo develops.
Empetraceae, 67
Empetrum, 67
Empress Tree, 109
ENDOSPERM: Food storage tissue of a seed derived from the triple-fusion nucleus of the embryo sac in the ovule. The term is sometimes loosely used to include perisperm, q.v.

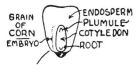

Figure 302

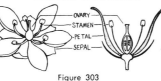

Figure 303

Figure 304

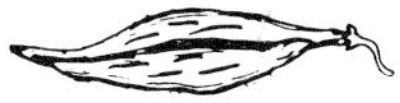

Figure 305

Grossulariaceae, 88
Guava, 86
Gunnera, 90
Gunnera family, 90
Gunneraceae, 90
Gymnosperms, 7, 10
GYNOBASE: An enlargement or prolongation of the receptacle of some flowers, as in the Boraginaceae and Lamiaceae.
GYNOBASIC STYLE: A style that is attached directly to the gynobase, as well as to the individual carpels or nutlets.

Figure 306

GYNOPHORE: A central stalk in some flowers, bearing the ovary.
Gypsophila, 49

H

Hackberry, 66
Haemodoraceae, 132
Halesia, 99
HALOPHYTE: A plant adapted to growth in salty soil.
Haloragaceae, 33
Hamamelidaceae, 38, 93
Hamamelis, 93
Harebell family, 113
hashish, 72
HASTATE: Shaped like an arrowhead, but with the basal lobes more divergent. (Compare sagittate.)
Hawaii state flower, 56
Haworthia, 132
Hawthorn, 87
Hazelnut, 40
HEAD: An inflorescence of sessile or subsessile flowers crowded closely together at the tip of a peduncle. Unless otherwise specified, a head is presumed to belong to the racemose group of inflorescences. (Compare sagittate.)
Heath, 96
Heath family, 96, 101
Hedera, 89
Hedychium, 134
Heliamphora, 16
Helianthemum, 48
Helianthus, 6, 112
Heliconia, 130
Heliconia family, 129
Heliconiaceae, 129

Heliotrope, 115
Heliotropium, 100
Helleborus, 28
Helxine, 71
HEMI: Greek prefix meaning half.
HEMITROPOUS OVULE: An ovule that is bent at about right angles to the stalk, so that it is intermediate between orthotropous and anatropous in shape.
Hemlock, 14
henna, 57
HERB: A plant, either annual, biennial, or perennial, with the stems dying back to the ground at the end of the growing season.
HERBACEOUS: Adjectival form of herb; also, leaflike in color or texture, or not woody.
Heuchera, 75
Hevea, 41
Hexastylis, 58, 85
Hibbertia, 25
Hibiscus, 56
Hickory, 3, 35
Hippocastanaceae, 79
Hippocratea family, 68
Hippocrateaceae, 68
Hippuridaceae, 32
Hippuris, 32
Hoary Puccoon, 100
Holly, 69
Holly family, 68
Hollyhock, 56
Holodiscus, 23
Honey Locust, 81
Honeysuckle, 114
Honeysuckle family, 114
Hop Goodenia, 112
Hop Hornbeam, 40
Hops, 72
Hops family, 71
Hoptree, 43
Horehound, 101
Horned Pondweed, 120
Horned Pondweed family, 119
Hornbeam, 40
Hornwort family, 32
Horse-Chestnut, 79
Horse-Chestnut family, 79
Horse Mint, 101
Horse-Radish Tree, 83
Horse-Radish Tree family, 83
Hosta, 133
Hound's-Tongue, 100
Houstonia, 113
Hoya, 103
Huckleberry, 96
Hudsonia, 48
Humulus, 72
Hyacinth, 133
Hydnocarpus, 47
Hydrangea, 87
Hydrangea family, 86
Hydrangeaceae, 86
Hydrocharitaceae, 115
Hydrocleys, 118
Hydrophyllaceae, 106
Hydrophyllum, 106
HYDROPHYTE: A plant adapted to life in the water.
HYPANTHIUM: A ring or saucer or cup around the ovary in some flowers, from which the sepals,

petals and stamens arise. When the petals and stamens appear to arise from the calyx tube, the hypanthium is that part of the apparent calyx tube which is below the attachment of the petals. In most flowers with a hypanthium, the ovary is free from the hypanthium and thus technically superior, and the flower is said to be perigynous. Some epigynous flowers (i.e., flowers with an inferior ovary) have a hypanthium prolonged beyond the top of the ovary.

Figure 307

Hypericum, 42
HYPO-: Greek prefix meaning beneath.
HYPOGYNOUS: With the perianth and stamens attached directly to the receptacle; more generally, beneath the ovary and ovaries. (Compare epigynous, perigynous.) See Fig. 335.

I

Icacina family, 65
Icacinaceae, 65
Idaho state flower, 87
Ilex, 69
Illiciaceae, 26
Illicium, 26
Illicium family, 26
IMBRICATE: Arranged in a tight spiral, so that the outermost member has both edges exposed, and at least the innermost member has both edges covered; more loosely, a shingled arrangement.
Impatiens, 78
IN- or IM-: Latin prefix meaning (in different contexts) not, in, or into.
INDEHISCENT: Not dehiscent, remaining closed at maturity.
Indian Almond family, 91
Indian Paintbrush, 110
Indian Pipe, 20
Indian Pipe family, 20
Indian Shot, 135
INFERIOR OVARY: An ovary with the other floral parts (calyx, corolla, and stamens) attached to its sum-

mit or attached to a hypanthium that is adnate to the ovary and projects beyond it. A flower with an inferior ovary is said to be epigynous.

Figure 308

INFLORESCENCE: A flower cluster of a plant, or, more correctly, the arrangement of the flowers on the axis.
INFRA-: Latin prefix meaning beneath, or within, or less than; opposite of supra-.
insectivorous plants, 15, 16, 17
INTEGUMENT: One of the one or two layers that form the outer covering of an ovule. The integument(s) develop into the seed coat. See Fig. 315.
INTER-: Latin prefix meaning between or among.
intercalary meristem, 126
INTERPETIOLAR STIPULES: Stipules that are attached to the stem in such a way as to connect the petioles of opposite leaves.
INTRA-: Latin prefix meaning within; opposite of extra-.
INTRASTAMINAL: Within the stamens, i.e., between the stamens and the ovary.
INTRUDED PLACENTA: A partial partition within an ovary, extending partway from the margin toward the center, and bearing ovules at its free edge.

Figure 309

INVOLUCEL: Diminutive of involucre; a secondary involucre. See Fig. 336.

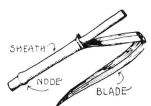

Figure 310

Figure 311

Figure 312

MERICARP: An individual carpel of a fruit that separates into its component carpels at maturity; one of the two parts into which the fruit of a member of the Apiaceae splits at maturity. See schizocarp.
-MEROUS: Greek combining form, referring to the number of parts in each circle of floral organs, usually with a numerical prefix, as trimerous, pentamerous, etc. Such terms often refer to the perianth only.

MESIC: Moist; neither very wet nor very dry.
MESOPHYTE: A plant adapted to growth under ordinary moisture conditions; intermediate between hydrophyte and xerophyte.

MICRO-: Greek prefix meaning small.
MICROPYLE: The opening through the integument(s) of the ovule to the nucellus. See Fig. 315.

MITRIFORM: Forming an elongate, hollow cap over some other structure, closed at the top.

MONAD: One of many individuals that are separate from each other rather than attached in groups.

MONO-: Greek prefix meaning one.
MONOCHLAMYDEOUS: Having only one set or series of perianth members. (Compare dichlamydeous.)
MONOCOTYLEDONS: A major group of angiosperms, in which the embryo typically has a single cotyledon.

MONOECIOUS: With unisexual flowers, both types borne on the same individual.

MONOTYPIC: Containing only a single type, as a genus with only one species.

MULTI-: Latin prefix meaning many.

MYCORHIZA: A symbiotic association of a fungus with the root of a vascular plant.

MYCOTROPHIC: Evidently modified as a result of mycorhizal association.

N

NAKED: Lacking various organs or appendages, in contrast to other forms in which these organs or appendages are present; e.g., a naked flower lacks a perianth.

NET-VEINED: With the veins forming a network. The term is traditionally applied to pinnate and palmate venation collectively, as opposed to parallel venation.

NODE: A place on a stem to which a leaf is (or has been) attached.

Figure 313

NUCELLUS: The inner covering of an ovule, surrounded by the integuments and enclosing the embryo sac.
NUDE: Same as naked.
NUMEROUS: As applied to floral parts, more than 10 in number.

NUT: A hard, dry, indehiscent, usually one-seeded fruit, larger and firmer-walled than an achene.

O

OCHREA: A sheath around the stem just above the base of a leaf, derived from stipules, as in many Polygonaceae.

ODD-PINNATE: Pinnately compound and with a terminal leaflet, so that typically there is an odd number of leaflets. (Compare even-pinnate.)

OPPOSITE: Situated directly across from each other at the same node, as the leaves of some plants; situated directly in front of (on the same radius as) another organ, as stamens opposite the petals. See Fig. 300.

ORTHOTROPOUS OVULE: A straight ovule, with the micropyle at the opposite end from the stalk.

OVARY: The structure that encloses the ovules of angiosperms; the expanded basal part of a pistil, containing the ovule(s).

Figure 314

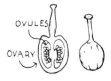

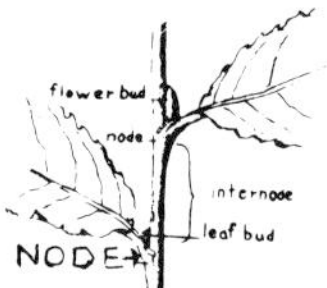

OVATE: Shaped like a long section through a hen's egg, with the larger end toward the base. (Term applied to plane surfaces; ovoid is the comparable term for a three-dimensional object.)

OVULE: A young or undeveloped seed; the megasporangium, plus the enclosing integuments, of a seed plant.

Figure 315

Oxalidaceae, 58
Oxalis, 58

P

Pachysandra, 42
Paeonia, 29
Paeoniaceae, 29
Palm family, 122
PALMATE: With three or more lobes or nerves or leaflets or branches arising from a common point. Fig. 316 shows a palmately veined leaf.

Figure 316

PALMATELY COMPOUND LEAF: A leaf with three or more leaflets arranged in palmate fashion. (Compare pinnately compound, trifoliolate.)

Figure 317

Palmetto, 122
Palms, 122

Panama Hat Plant, 122
Panax, 89
Pandanaceae, 123
Pandanus, 123
PANICLE: A branching inflorescence with a central axis and lateral branches; strictly, such an inflorescence, with the sequence of blooming proceeding uniformly from the base to the top.

Figure 318

Papaver, 53
Papaveraceae, 45, 53
Papaya, 98
Papaya family, 98
PAPILIONACEOUS FLOWER: A flower having the structure typical of the Fabaceae, with a banner petal, two wing petals, and two partly connate keel petals.
PAPPUS: The modified calyx crowning the ovary (and achene) of the Asteraceae, consisting variously of hairs, scales, bristles, or a mixture of these.
Para Rubber, 41
PARALLEL-VEINED: With several or many more or less parallel main veins, the network of smaller veins not obvious. (Compare net-veined.)

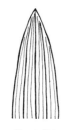

Figure 319

PARASITE: A plant that grows attached to some other living plant, obtaining nourishment from it.
parasitic plants, 17-20, 40, 80, 110
PARIETAL PLACENTA: A placenta along the walls or on the intruded partial

partitions of a compound, unilocular ovary. (Compare marginal, axile, and free-central placenta.)

Figure 320

Parnassia, 75
Parnassiaceae, 75
Parrot's Feather, 34
Parsley, 90
Parsnip, 90
Parthenocissus, 64
Passiflora, 74
Passifloraceae, 74, 101
Passion-Flower family, 74, 101
Pastinaca, 90
Paulownia, 109
Pawpaw, 26
Pears, 87
Pea family, 80
Peas, 81
Pedaliaceae, 109
PEDICEL: The stalk of a flower in an inflorescence.

Figure 321

Pedicularis, 110
PEDUNCLE: The stalk of an inflorescence or of a single flower.
Pelargonium, 59, 78
PELTATE: Shield-shaped, attached by the lower surface instead of by the base or margin.

Figure 322

Pennsylvania state flower, 96
Penstemon, 110
Peony, 29
Peony family, 29
Peperomia, 39
pepper, 39
Pepper family, 38

Pepper-tree, 48
PERFECT FLOWER: A flower with one or more stamens *and* one or more pistils, whether or not it has a calyx and corolla. (Compare unisexual flower.)
PERIANTH: All of the sepals and petals (or tepals) or a flower, collectively.

Figure 323

PERIGYNOUS: With a superior ovary, surrounded by a hypanthium (q.v.) to which the sepals, petals (if present) and stamens are attached; more generally, around the base of the ovary (or the ovaries collectively), as a perigynous disk. See Fig. 307.
PERISPERM: Food storage tissue in the seed, derived from the nucellus of the ovule rather than from the triple fusion nucleus.
Periwinkle, 103
Persea, 44
Persimmon, 98
PERSISTENT: Remaining attached after the normal function has been completed.
PETAL: A member of the second set of floral leaves (i.e., the set just internal to the sepals), usually colored or white and serving to attract pollinators.

Figure 324

PETALOID: Petal-like, especially in color and texture.
PETIOLE: A leaf stalk. See Figs. 316, 317.
Phacelia, 106
Phellodendron, 43
Philadelphus, 87
Philodendron, 123
Phlox, 102
Phlox family, 102
Pholisma, 20
PHOTOSYNTHETIC: Capable of manufacturing food from raw materials, with light as the source of energy; the green pig-

ment chlorophyll is one of the essential enzymes for photosynthesis.

PINNATE: With two rows of lateral appendages, or parts along an axis, like the barbs on a feather. Fig. 325 shows a pinnately veined leaf.

Figure 325

PINNATELY COMPOUND LEAF: A leaf with 3 or more leaflets arranged in pinnate fashion. (Compare palmately compound, trifoliolate.)

Figure 326

PISTIL: The female organ of a flower, composed of one or more carpels, and ordinarily differentiated into ovary, style, and stigma. A flower may have one or more pistils;

when there is more than one pistil, the pistils are always simple, i.e., each composed of a single carpel. See Fig. 296.

PISTILLATE FLOWER: A flower with one or more pistils, but no stamens.

Figure 327

PITH: A central, usually soft tissue in a stem (or seldom in a root), surrounded by vascular bundles or a ring of vascular tissue.

PLACENTA: The tissue of the ovary to which the ovules are attached. For placentation types, see axile, free-central, marginal, and parietal placentas, Figs. 291, 292, 305, 312.

POLLEN GRAIN: One of the numerous tiny structures, produced in a pollen sac, and within which sperms are eventually formed.

POLLEN SAC: One of usually two similar units of an anther, containing pollen grains.

POLLINATION: In angiosperms, the transfer of pollen from the anther to the stigma; in gymnosperms, from the microsporangium to the micropyle.

POLLINIUM: A coherent cluster of numerous pollen grains.

POLY-: Greek prefix meaning many.

POLYPETALOUS: With the petals separate from each other. (Compare sympetalous.)

PROXIMAL: At the near end. (Compare distal.)

Q

R

RACEME: A more or less elongate inflorescence with

flowers on pedicels arising in succession from the bottom upwards along an unbranched axis.

Figure 328

RACEMOSE INFLORESCENCE: One of a general class of inflorescences characterized by blossoming from the bottom upwards, or from the outside toward the center. (Compare cyme.)

RAPHE: The part of the funiculus that is permanently adnate to the integument of the ovule, commonly visible as a line or ridge on the mature seed coat.

RECEPTACLE: The end of the stem (pedicel), to which the other flower parts are attached.

REGULAR FLOWER: A flower in which the members of each circle of parts (or at least the sepals and petals) are similar in size, shape, and orientation.

RENIFORM: Kidney-shaped.

REPLUM: A persistent, framelike placenta that bears ovules along the margins, as in the Brassicaceae and Capparaceae. In the Brassicaceae the replum forms a thin partition that divides the ovary into two locules, but the ovules are still attached along the margins. See Figs. 139, 140.

Figure 329

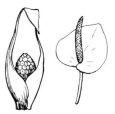

Figure 330

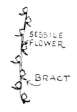

Figure 331

Spinacia, 73
Spiraea, 23
Spirodela, 115
Spring Beauty, 70
Spruce, 14
Spurge family, 34, 41
Squash, 111
Squash family, 111
STAMEN: The male (pollen-bearing) organ of a flower, usually consisting of an anther (q.v.) and a filament (q.v.). See Fig. 290.
STAMINATE FLOWER: A flower with one or more stamens but no pistil.

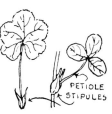

Figure 332

STAMINODE: A modified stamen, which does not produce pollen.
Stapelia, 103
Staphylea, 67
Staphyleaceae, 66
Star Apple, 98
Star Cucumber, 111
STELLATE: Star-like. Stellate hairs have several or many branches from the base.
Stemona family, 130
Stemonaceae, 130
Sterculia, 56
Sterculiaceae, 56, 63
Stewartia, 54
STIGMA: The part of a pistil that is receptive to pollen; the stigma is usually elevated above the ovary on a style.

Figure 333

Stinging Nettle, 71
STIPE: The stalk of a structure, without regard to its morphological nature. The term is usually applied only where more precise terms such as petiole, pedicel, or peduncle cannot be used, as the stipe of an ovary.
STIPITATE: Borne on a stipe.
STIPULE: One of a pair of

basal appendages found on many leaves.

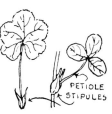

Figure 334

Stone-Plant, 53
Stonecrop family, 28
Storax family, 98
Stork's Bill, 76
Strawberry, 30
Strawberry Shrub, 24
Strelitziaceae, 129
Striga, 19
Strophanthus, 103
strychnine, 104
Strychnos, 104
STYLE: The slender stalk that typically connects the stigma(s) to the ovary. See Figs. 296, 333.
Styracaceae, 98
Styrax, 99
Subularia, 8
SUCCULENT: Fleshy and juicy; more specifically, a plant that accumulates reserves of water in the fleshy stems or leaves, due largely to the high proportion of hydrophilic (water-attracting) colloids in the protoplasm and cell sap. Succulents frequently have a special twist on photosynthesis, called crassulacean acid metabolism, which permits them to absorb carbon dioxide at night and keep their stomates closed during the day.
Sugar Maple, 43
Sumac family, 48, 64
Sundew family, 16
SUPER- SUPRA-: Latin combining forms meaning above, or upon, or more than; opposite of infra-.
SUPERIOR OVARY: An ovary that is attached at the summit or center of the receptacle and is free from all other flower parts. A flower with a superior ovary may be either hypogynous or perigynous.

Figure 335

SUPERPOSED: Placed one above or on top of another.
Suriana, 24
Surianaceae, 24
SUTURE: A seam or line of fusion; usually applied to the vertical lines along which a fruit may dehisce.
Sweet Bay, 25
Sweet Gale, 39
Sweet-Leaf, 99
Sweet Potato, 106, 131
Sweet William, 49
Swietenia, 62
SYMBIOSIS: A close physical association between two different kinds of organisms, typically with benefit to both.
Sycamore family, 22
SYMPETALOUS: With the petals connate, at least toward the base. (Compare polypetalous.)
Symphoricarpos, 114
Symplocaceae, 99
Symplocarpus, 123
Symplocos, 99
Symplocos family, 99
Syngonanthus, 125
Syringa, 87
Syringa, 107
Syringodium, 119
Syzygium, 86

T

Talinum, 49
Tamaricaceae, 44
Tamarisk, 44
Tamarisk family, 44
Tamarix, 44
Tape-grass, 116
Taraxacum, 112
Taxaceae, 13
Taxodiaceae, 14
Taxodium, 15
taxonomic patterns, 5
taxonomic rank, 2, 3
TAXONOMY: A study aimed at producing a system of classification of organisms that best reflects the totality of their similarities and differences.
Taxus, 13
Tea family, 54
Teak wood, 100
Teasel, 114
Teasel family, 114
Tectona, 100
Telesonix, 57
Tennessee state flower, 133
TEPAL: A sepal or petal, or a member of an undifferentiated perianth. The term is used especially

when it is difficult to be sure whether the structures are sepals or petals, or where the sepals and petals are similar.
Terminalia, 91
TERNATE: In threes.
TERRESTRIAL: Growing or living on land.
TETRA-: Greek prefix meaning four.
TETRAD: A group of four.
TETRADYNAMOUS: With four long and two short stamens, as in many Brassicaceae.
Texas state flower, 81
THALLOID: Resembling or consisting of a thallus.
THALLUS: A plant body that is not clearly differentiated into roots, stems, and leaves.
Thea, 54
Theaceae, 54
Theobroma, 56
Theophrasta family, 95
Theophrastaceae, 95
Thread Palm, 122
Thuja, 14
Thyme, 101
Thymelaeaceae, 51
Thymus, 101
Tiarella, 46
Tilia, 56
Tiliaceae, 56
Tillandsia, 128
Tobacco, 107
Tomato, 107
Touch-Me-Not, 78
Touch-Me-Not family, 77
Toxicodendron, 65
Tradescantia, 128
Trailing Arbutus, 96
Trapa, 31
Trapaceae, 31
Traveler's Tree, 129
Tree-of-Heaven, 62
TRI- TRIPLO-: Latin or Greek prefix meaning three.
Triadenum, 42
Tribulus, 60
TRIFID: Cleft into three segments.
TRIFOLIOLATE LEAF: A leaf with three leaflets. A trifoliolate leaf may be pinnately or palmately trifoliolate, depending on the position of the terminal leaflet with respect to the lateral ones. Poison Ivy (Fig. 127a) has pinnately trifoliolate leaves, and White Clover (Fig. 167b) has palmately trifoliolate leaves.
Trifolium, 58, 81
Triglochin, 121
Trillium, 127
TRILOCULAR: With three locules.
TRIMEROUS: With three parts of a kind.
Triodanis, 113
Tropaeolaceae, 77
Tropaeolum, 77
Trumpet-Creeper, 109
Trumpet-Creeper family, 108

Tsuga, 14
Tuberous Begonia, 86
Tulip Poplar, 3
Tulip Tree, 25
Tulips, 133
Tumbleweed, 73
Turk's-cap Lily, 133
Turnera, 75
Turnera family, 74
Turneraceae, 74
Typha, 121
Typhaceae, 121

U

Ulmaceae, 65
Ulmus, 66
UMBEL: An inflorescence of the racemose type with a very short axis and more elongate pedicels which seem to arise from a common point. In a compound umbel the primary branches are again umbellately branched.

Figure 336

Umbellularia, 44
UNI-: Latin prefix meaning one.
Unicorn Plant, 109
UNILOCULAR: With a single locule.
UNISEXUAL FLOWER: A flower with one or more stamens, or with one or more pistils, but not both.
Urtica, 71
Urticaceae, 71
Utah state flower, 127
Utricularia, 15

V

Vaccinium, 96
Valerian family, 114

Valeriana, 115
Valerianaceae, 114
Vallisneria, 116
VALVATE: Arranged with the margins of the petals (or sepals) adjacent throughout their length, without overlapping; opening by valves.
Vanilla Leaf, 45
VASCULAR: Pertaining to conduction. Vascular plants are those which have xylem (a water-conducting tissue) and phloem (a food-conducting tissue); a vascular bundle is a strand of xylem and phloem and associated tissues.
VENTRAL: Pertaining to or located on the front or belly side. The adaxial side of a leaf or carpel is considered to be the ventral side, and the seed-bearing suture of an ordinary carpel is the ventral suture. (Compare dorsal.)
Venus' Flytrap, 16
Venus' Locking-Glass, 113
Verbena, 100
Verbena family, 100, 110
Verbenaceae, 6, 100, 110
Vermont state flower, 81
Veronica, 110
Vertebrates, 5
VESSEL: A specialized water-conducting element in the xylem, consisting of several cells placed end to end, without cross-walls or with the cross-walls perforated.
VESTIGIAL: Much reduced and hardly or not at all functional. The term implies that the structure was better developed in ancestral forms. (Compare rudimentary.)
Vetch, 81
Viburnum, 114
Vicia, 81
Victoria, 31
Vinca, 103

Viola, 83
Violaceae, 83
Violet, 83
Violet family, 83
Virginia Creeper, 64
Virginia Heart-Leaf, 58
Virginia state flower, 93
Viscaceae, 18
Vitaceae, 63
Vitis, 64

W

Wake-Robin, 127
Walnut family, 35
Wartweed, 35
Washington state flower, 96
Washingtonia, 122
Water Chestnut, 32
Water Chestnut family, 31
Water Hemlock, 90
Water Hyacinth, 117
Water Lily family, 31
Water Milfoil family, 33
Water Nymph, 119
Water-Nymph family, 119
Water Plantain, 117
Water Plantain family, 117
Water Poppy, 118
Water-poppy family, 118
Water Purslane, 32
Water Smartweed, 33
Water Starwort family, 34
Water-weed, 116
Waterleaf, 106
Waterleaf family, 106
Watermelon, 111
Waterwort family, 33, 76
Wedeliella, 80
Welwitschia family, 12
Welwitschiaceae, 11, 12
West Virginia state flower, 96
Western Red Cedar, 14
Wheat, 126
White Alder, 60
White Alder family, 60
White Ash, 36
White Clover, 58
White Pine, 14
White Spruce, 14
WHORL: A ring of three or more similar structures radiating from a node or common point. Fig. 337 shows whorled leaves.

Figure 337

Wild Buckwheat, 50
Wild Ginger, 85
Wild Plantain, 130
Wild Yam, 131
Willow family, 38
WING: A thin, flat extension or projection from the side or tip of a structure; one of the two lateral petals in a flower of the Fabaceae.
Winteraceae, 26
Wintergreen, 48
Wisteria, 81
Witch Hazel, 93
Witch Hazel family, 38, 93
Wood Sorrel family, 58
Woolly Pipe-Vine, 85
Wyoming state flower, 110

X

XERO-: Greek prefix meaning dry.
XEROPHYTE: A plant adapted to life in dry places.
Ximenia, 40
Xyridaceae, 127
Xyris, 128

Y

Yam family, 131
Yams, 131
Yellow-eyed Grass family, 127
Yerba Mansa, 91
Yew, 13
Yew family, 13
Yucca, 132

Z

Zamia, 11
Zannichellia, 120
Zannichelliaceae, 119
Zanthoxylum
Zingiberaceae, 134
Zinnia, 112
Zostera, 119
Zosteraceae, 119
Zygophyllaceae, 59